Praise for *The Arrogant Ape*

"A gifted primatologist, Christine Webb undertakes a profound takedown of the human supremacy complex. She accomplishes this not by arguing that our species is unexceptional, but by showing that *every* species is marvelously exceptional—each creature enacting its own uncannily weird way of relating to the rest. *The Arrogant Ape* is not only lucidly thought and eloquently articulated; it's a deeply *felt* work, empathic and curious and wonderstruck. It opens us to a more humble and collaborative natural science—one that engages nature not as a quantifiable set of mechanical objects waiting to be figured out and mastered, but as a collective of often-incommensurable yet oddly intersecting styles of sentience, a breathing community of *subjects* to whom we can only apprentice ourselves."

—DAVID ABRAM, AUTHOR OF *BECOMING ANIMAL*

"*The Arrogant Ape* convincingly argues that humans need not—even should not—be placed central to every discussion in science, in policy, or in considering how to live one's life. This book will reframe the way you view animal welfare, and may even reframe the way you see."

—ALEXANDRA HOROWITZ, AUTHOR OF *INSIDE OF A DOG*

"Webb makes clear that the notion that we're the most important show in town—smarter than, better than, more important than, uniquely exceptional, above, and separate from other animals—has got it all wrong. This distorted view of humans in which we use ourselves as some sort of standard to which individuals of other species should strive is not only arrogant, but singularly ill-informed. I highly recommend *The Arrogant Ape*."

—MARC BEKOFF, AUTHOR OF *THE EMOTIONAL LIVES OF ANIMALS*

"Webb offers a deeply considered, self-reflective, and undeniably philosophical approach to the scientific study of animal behavior. She is spearheading a paradigm shift in science, deftly folding in Indigenous and phenomenological perspectives to forge a hybrid approach to empirical knowledge-seeking. Her book is a modern exploration of the ancient speciesism problem, leading the reader toward a hopeful appeal that we can dispel our culturally acquired forms of anthropocentrism in service of a humbler path to understanding both the animal mind and humanity's connection to the natural world."

—JUSTIN GREGG, AUTHOR OF *IF NIETZSCHE WERE A NARWHAL*

"A brilliant and moving invitation to rethink the illusory human-animal divide. In these vivid stories and incisive reflections, we come to see not only how blinkered we have often been but, more importantly, how we can expand experience, deepen science, and open imaginations when it comes to our relationships with other animals. A must read for all who love the more-than-human world."

—DAVID GEORGE HASKELL, AUTHOR OF *SOUNDS WILD AND BROKEN*

"Webb puts us in our place (and a fine place that is, in fact), showing us how much more fascinating the world is if we see it as it is, rather than denigrating it and using it as a mere resource."

—CHARLES FOSTER, AUTHOR OF *CRY OF THE WILD*

"We humans have a favorite story: we are different, we're the best; in this world, we matter most. With insights gained through a lifetime of intimately observing our primate relatives, Webb offers us a different reflection when we look in the mirror, and charts for us a saner path for nature and ourselves."

—CARL SAFINA, AUTHOR OF *ALFIE AND ME*

"A brilliant book filled with insights into the living world we inhabit of diverse intelligences, forms of language, sensory feats, and entangled relations. Webb makes a powerful case against the bias of anthropocentrism that blinds us to Earth's reality by making 'man the measure of all things.' In choosing to relinquish this delusion we become more virtuous, we acquire more valid knowledge about nature, and we give ourselves the chance to live in graceful coexistence with all life. A must read on matters of critical importance at this pivotal moment for humanity and Earth."

—EILEEN CRIST, AUTHOR OF *ABUNDANT EARTH*

"Hubris" by Mary Jane Webb (2020)

The
Arrogant Ape

The Myth of Human Exceptionalism and Why It Matters

Christine Webb

Avery
an imprint of Penguin Random House
New York

AVERY

an imprint of Penguin Random House LLC
1745 Broadway, New York, NY 10019
penguinrandomhouse.com

Avery with colophon is a trademark of Penguin Random House LLC

Most Avery books are available at a discount when purchased in quantity for sales promotions
or corporate use. Special editions, which include personalized covers, excerpts, and corporate
imprints, can be created when purchased in large quantities. For more information,
please e-mail specialmarkets@penguinrandomhouse.com. Your local bookstore can also
assist with discounted bulk purchases using the Penguin Random House corporate
Business-to-Business program. For assistance in locating a participating retailer,
e-mail B2B@penguinrandomhouse.com.

Book design by Angie Boutin

Library of Congress Cataloging-in-Publication Data

Names: Webb, Christine, author.
Title: The arrogant ape: the myth of human exceptionalism and
why it matters / Christine Webb.
Description: New York: Avery, [2025] | Includes index. |
Identifiers: LCCN 2024050380 (print) | LCCN 2024050381 (ebook) |
ISBN 9780593543139 (hardcover) | ISBN 9780593543153 (epub)
Subjects: LCSH: Human evolution. | Exceptionalism—Social aspects. | Speciesism.
Classification: LCC GN281 .W43 2025 (print) | LCC GN281 (ebook) |
DDC 304.2—dc23/eng/20250220
LC record available at https://lccn.loc.gov/2024050380
LC ebook record available at https://lccn.loc.gov/2024050381

ISBN 9798217178711 (International edition)

Printed in the United States of America
1st Printing

The authorized representative in the EU for product safety and compliance is
Penguin Random House Ireland, Morrison Chambers, 32 Nassau Street,
Dublin D02 YH68, Ireland, https://eu-contact.penguin.ie.

To Frans de Waal,
my dear mentor and a most humble ape

CONTENTS

CHAPTER ONE

The Human Superiority Complex

W HAT A PIECE OF WORK IS MAN!" MARVELS HAMLET, "HOW
noble in reason! how infinite in faculty! . . . in action, how
like an angel! in apprehension, how like a god! . . . the paragon of
animals!"

In a few short lines, Shakespeare gives us the most prominent
theme in the history of Western thought: human beings are the most
clever, moral, and capable species on earth.

But I wonder, if we truly believe we are so much better than other
species, why have we spent thousands of years driving home the
point?

Psychologists have shown that people overemphasize their own
abilities and accomplishments to conceal actual feelings of short-
coming and failure. When it comes to other species, do we have a
so-called superiority complex?

After all, we're not the biggest, fastest, or strongest. Blue whales,
cheetahs, and rhinoceros beetles outdo us there. Nor are we the most
numerous or long-lived. Ants and sea sponges (not to mention most
bacteria and plant species) easily have us beat there too. Other species

outperform humans in countless ways. Just try competing with ea-gles on vision tests or dolphins on tests of echolocation. So we turned to our intellect. It was decided! We must have the edge there.

Carl Linnaeus, the father of the modern system for classifying organisms, named us *Homo sapiens*, the "wise man." Today we call ourselves *Homo sapiens sapiens*, the wisest of the wise. Seeing ourselves in superlatives flies in the face of Darwinian notions of continuity between species.

Yet even today, examples of human exceptionalism by high-profile media outlets and scholars are rampant. Just notice how of-ten the human species is ranked as a separate, superior entity to the environment, either directly or by implication. "Although we are animals . . . we are not *just* animals," writes philosopher Roger Scru-ton in a 2017 *New York Times* op-ed. "As persons, we inhabit a life-world that is not reducible to the world of nature." In a 2018 article in *The Guardian* titled "The Human League: What Separates Us from Other Animals?," geneticist Adam Rutherford asserts, "We cannot satisfactorily match our behaviour with other beasts, and claims that we can are often poor science." As comparative psycholo-gist Thomas Suddendorf avows in a piece for CNN, "It seems obvi-ous that there is something extra special about us."

The quest to identify humanity's uniquely distinguishing trait is also teeming with human exceptionalism. In a 2016 NPR interview titled "Why Did Humans Become the Most Successful Species on Earth?," historian and bestselling author Yuval Harari credits the hu-man imagination for "why we can cooperate in our billions, whereas chimpanzees cannot, and why we have reached the moon and split the atom and deciphered DNA, and they just play with sticks and bananas." A *Los Angeles Times* article instead cites our unique gener-osity: "Maybe the reason we call it 'human kindness' is because that's the only kind there is." The general-interest publishing landscape is saturated with titles proclaiming to have discovered that elusive holy grail of what makes us *human*, touting our species' unrivaled intelligence, adaptability, friendliness, language, creativity, and the like.

Human exceptionalism is not restricted to the popular press. There are entire conferences on what makes humans special. At the 2019 Association for Psychological Science convention, primatologist Michael Tomasello opened his keynote address with an apology to great apes before making the case for our cognitive superiority as humans. There's an entire research enterprise on this vast mental rift between humans and other animals, dubbed "humaniqueness" by biological anthropologist Marc Hauser. The running story is that while other primates possess the rudimentary building blocks of human cognition, they lack the unique adaptations of the human mind that allowed our species to thrive and flourish. These traits have enabled humans to dominate the planet—a dominance often equated to our "evolutionary success." What else would you expect from a species that frequently flaunts its own intelligence? It seems as much today as historically, what makes us special is that we like to think it so. Hamlet got one thing right: we're a piece of work.

The argument of *The Arrogant Ape* is that human exceptionalism—a.k.a. anthropocentrism or human supremacy—is at the root of the ecological crisis. This pervasive mindset gives humans a sense of dominion over Nature, set apart from and entitled to commodify the earth and other species for our own exclusive benefit. And it's backfiring on us today, spurring forest fires, sea level rise, mass extinctions, and pandemics like the coronavirus.

This unfortunate and dangerous way of viewing our world is a brainwashing of such major proportion that many people remain entirely unaware of it. From a young age, the myth of human exceptionalism gets internalized and reinforced by society in various ways—in schools and textbooks, sermons, political campaigns, advertising, movies, holiday celebrations, language, and more. But perhaps most unsettlingly, human exceptionalism has even seeped into our sciences.

Having spent my career studying the minds of our closest living primate relatives, I know this firsthand.

One morning, on the edge of the Namib Desert, a baboon named Bear read my mind. The day before, he and a dozen other baboons had mobbed my colleague, emitting loud, high-pitched barks and slapping her legs. We weren't sure what had caused it, but Bear was the clear instigator. That following morning, I was keeping my distance from the troop, worried about a repeat incident, when over the ridge marched Bear and his entourage again. This time, they made a beeline for me. I was wobbling down a steep, rocky incline and could not get out of their way. My heart was racing and my palms were sweaty, but outwardly, I remained calm. Then something happened that I will never forget, because it changed my entire view of other animals and what they are capable of. Despite my tranquil act, Bear approached and put his hand on my leg. He looked up at me and bared his teeth in an awkward, forced grimace. As a primatologist, I know this is a pacifying gesture, something baboons do to avert and resolve conflicts with each other. He was making amends for the day before. This baboon knew what I knew and was trying to put things right.

Why was this encounter so special? Reflecting on it later that day from the safety of my tent, I recalled all the times I had been told as a graduate student that other primates lack mind-reading abilities— what scientists call a "theory of mind." This is one of many cognitive capacities said to set humans apart from other animals. Baboons aren't supposed to know what other baboons know, let alone what a member of another species might be thinking. But Bear? Bear read my mind. I've had to unlearn a lot of things that were stated as fact when I was a student.

For centuries, Western philosophy and religion sustained the belief that the human species occupies a central, superior place in the universe. Charles Darwin revolutionized this worldview with his theory of evolution by natural selection, showing that species form an interconnected tree of life rather than a hierarchy. Darwin himself would likely be surprised by human exceptionalism's lasting impact on the collective imagination. But he knew his ideas threatened a society bent on maintaining this myth. As he wrote in one of his notebooks more than twenty years before publishing *On the Origin of Species*: "Man in his arrogance thinks himself a great work, worthy the interposition of a deity. More humble, and I believe truer, to consider him created from animals."

Today, science is increasingly responsible for shaping our understanding of humanity's position in the natural world. But when we let the ideology of human supremacy infiltrate science, it leads to biases that perpetuate human exceptionalism rather than a more humble, authentic view of human capabilities. This is one of the main reasons why the myth of human exceptionalism dominates our thinking today. This is why—in contrast to Darwin's theory of mental continuity between humans and other species—contemporary scholars maintain that a profound gap separates our mind from the animal kind.

The idea for *The Arrogant Ape* began in 2019 when I joined Harvard's Department of Human Evolutionary Biology. How humans evolved to be the way they are is a question that has long interested me. Beyond a lifelong curiosity about other animals, it is part of the reason I became a primatologist. An evolutionary perspective emphasizes continuity—differences between species that are a matter of *degree* rather than *kind*. And yet, the idea that there are essential traits enjoyed by all and only members of our species (or any species, for that matter) has been incredibly resilient. Historically, attempts to define some stable, unique, and universal "human nature" have either included members of other species or excluded an array of humans (often those already discriminated against and marginalized in some way by our society).

But as long as we stress a narrative of human uniqueness, we might as well do the same for other lifeforms. All species have evolved specialized adaptations to their environments. If humans are unique, then every species is unique. However, human exceptionalism is different from human uniqueness. Human exceptionalism suggests that what is distinctive about humans is more worthy and advanced than the distinguishing features of other forms of life.

One might assume that human exceptionalism has been laid to rest with the widespread scientific acceptance of evolution. But this worldview is so deeply embedded in our culture that, often without conscious effort, almost everyone (including scientists) accepts its basic premises.

In the following chapters, we will engage with ideas that sometimes run counter to much of what we ordinarily consider true about the world—assumptions so inherent that we may be unaware we're even assuming anything. Human exceptionalism has been repeated and left unchallenged to the point that we hardly recognize it as a story anymore; instead, we've internalized it as "fact." But unlearning this worldview can be as rewarding as it is challenging. We (and many other species) stand to gain so much when we become more conscious of our own biases. This is a book about how learning to recognize the pervasiveness of human exceptionalism can change how we look at the world, and has changed how many people, including myself, think about and do science.

From the arid deserts of Namibia where an extraordinary baboon troop thrives to a forested sanctuary in Zambia where chimpanzees are rescued and rehabilitated, I have spent most of my adult life researching the rich social, emotional, and cognitive lives of nonhuman primates. They have taught me many things. But above all, they have taught me that the lines we think separate humans from other species are artificial, because the ways we go about drawing them are fundamentally flawed.

For instance, most claims of humans' cognitive uniqueness are based on experiments that compare the abilities of captive chimpanzees with those of fully autonomous Western humans. The overwhelming conclusion of these studies is that humans clearly outperform apes in various cognitive domains including theory of mind, cooperation, altruism, metacognition, joint attention, and prosociality. But too often the deck is stacked against other species, and the development of hypotheses, the design of experiments, and the evaluation of evidence are skewed in humans' favor.

We assume that these caged chimpanzees and free human populations are representative of each species, when they are not. The chimps have typically spent their entire lives isolated with a small group in restricted, human-made conditions. I have also studied nonhuman primates in such settings—labs, zoos, and sanctuaries. Captive chimpanzees are nothing like their wild counterparts. The human groups are also unrepresentative of humanity on the whole: recent research tells us that they are among the most psychologically unusual in the world (Western, educated, industrialized, rich, and democratic—WEIRD). Thus, the comparison reveals very little about the different cognitive abilities of these two species.

Moreover, these studies rely on human-centric experimental designs. They involve tasks that other species would never naturally encounter, such as computer touch screens and plastic toys. This kind of research can only tell us how other species perform on tasks in which humans excel. It tells us very little about other species' own evolved cognitive adaptations. It would be like presenting WEIRD human participants with sticks, rocks, and nuts of varying sizes and measuring their intelligence by comparing their performance with that of chimps on termite-fishing or nut-cracking tasks—tasks that involve mental foresight, manual dexterity, sustained attention, and causal reasoning. Would we conclude that humans are inferior to chimpanzees in said cognitive capacities on the basis of their performance? The satirical online publication *The Onion* summed up this anthropocentric bias rather well in an article titled "Study: Dolphins Not So Intelligent on Land." When you measure the world with a ruler made for humans, other species will inevitably come up short.

A less human-centric paradigm would bring us a long way toward understanding other species' own evolved cognitive adaptations, rather than comparing them to a human standard that inevitably renders them deficient. Pioneering scientists past and present have broken from the pressures and limitations of human exceptionalist thinking. Their work looms large in my own research. From household names like Charles Darwin to lesser-known visionaries like Lynn Margulis, from botanists like Robin Wall Kimmerer to primatologists like Frans de Waal, what happens when scientists approach their study systems with humility, reverence, and an open mind? Their discoveries reveal underappreciated complexities of nonhuman life—from the languages of songbirds and prairie dogs, to the cultures of chimpanzees and reef fishes, to the acumen of plants and fungi. A different way of looking at organisms, one that is possible if we overcome notions of human exceptionalism and think about species on their own terms, is revolutionizing our perception of them and of ourselves. The research becomes a powerful metaphor for ways of understanding and living in the world—ways that Indigenous cultures have long modeled and protected. This book argues that this less human-centric approach is both possible and necessary. It is key to better science and a richer, sustainable way of life.

Critiques of human exceptionalism tend to focus on our moral obligation toward other species. What they overlook is what humanity also stands to *gain* by dismantling its illusions of uniqueness and superiority. And not just because these illusions are at the root of the environmental crisis. But because they prevent us from engaging with the world in a way that continuously instills a sense of awe, wonder, and humility. When we are not blinded by a human-centric lens, we start to feel more like the integrated part of Nature that we are.

I teach an undergraduate course called The Arrogant Ape. I witness my students undergo major transformations as they learn to see past the basic ways their sense of the world has been framed by human exceptionalism. As the wool is pulled away from their eyes, they come to experience Nature as more alive, animate, and aware. A walk through campus or the woods is never the same; it's an opportunity to interact with a multitude of other lifeforms, to feel a part of something greater than the self. For some of us, unlearning human exceptionalism reinforces what we've long held to be self-evident: that the world is filled with diverse kinds of intelligence and awareness. But for others, the experience is more like a reawakening—remembering a childlike curiosity for and connection with the living world. My students' experiences and my own have encouraged me to collect these ideas together in the book you are reading.

This renewed relationship restores us. It rejuvenates us. It satisfies one of our oldest and deepest desires to *belong* to the larger whole we inhabit. And in turn, it empowers us toward actionable change. For my students, unlearning human exceptionalism ignites an ecological consciousness that many redirect into environmental or animal advocacy. When you perceive the world as an object, its destruction becomes meaningless. But when you understand that the world is an animate entity of which you are a part, activism isn't a choice; it be-comes a way of life. That's thanks to one simple but often overlooked truth: what people do about Nature depends on how they see them-selves in relation to Nature. Once delusions of human superiority and separateness are broken down, we can no longer passively watch Nature's destruction, in part because we see what we stand to gain— not merely in the end but in the *here and now*. This is a refreshing departure from the dominant environmental narrative, which em-phasizes sacrifice, costs, and long-term adverse impacts. Nature is not a means to human ends but an interdependent system whose well-being ultimately determines our own.

The coronavirus pandemic provides a timely illustration. Nature seems to be defying the primacy of humans like never before. And yet, media coverage praises humans' ingenuity for creating vaccines, ignoring the fact that human exploitation of animal habitats possibly triggered the virus in the first place (and will inevitably lead to future zoonotic outbreaks). At the same time, militarized discourses of a "war on" or "fight against" the virus perpetuate the view that Nature is a force to be controlled and dominated. Such human-centered nar-ratives also figure in recent discussions about climate change and en-vironmental "techno-fixes" like solar geoengineering and colonizing Mars. These framings bypass a critical opportunity to deconstruct forms of human exceptionalism in the public imagination. They in-stead assure us that mankind will prevail over Nature after all. But this same narrative of human progress and mastery got us into this mess, and it would be wise to recognize that we can no longer rely on the values, institutions, and scientific methods that gave rise to this moment to get us out of it. We need a radically humbler approach. And the clock is ticking.

I am running with the assumption that you agree that planetary health is in a dire situation. In this book, I am not trying to con-vince anyone that such a situation exists. The statistics alone are harrowing—we're likely all haunted by some version of them. More than 90 percent of the earth's soil will be degraded by 2050, based on current trends. Thirty percent of global forest cover has been cleared, another 20 percent is deteriorated, and most of the rest is now fragmented, leaving only about 15 percent intact. Earth's temperature has risen by about 2 degrees Fahrenheit since 1950, and the rate of warming since 1982 has more than tripled. Ocean acidification has been occurring one hundred times faster than what has occurred during natural events over the past fifty-five million years. Wild animal populations plummeted by nearly 70 percent over the last half century. Pollinators on whom humans and many other species depend are facing extinction. As I write, these beings are vanishing from the planet and from our daily lives.

"I could go on," writes the recovering environmentalist Paul Kingsnorth after a barrage of such statistics, "but I suspect you've heard it all before and that, like the rest of us, you have no idea what to do about it, or whether anything can be done at all."

Why is it that we have not taken the steps needed to address the devastating consequences of climate change (which imperil our very own survival) when faced with a scientific consensus that it demands urgent action? After all, we've known the "facts" for a long time (in the 1960s, scientists were concerned enough about human-caused climate change to formally warn U.S. president Lyndon B. Johnson). These days, the obstacles are hardly physical or technological. This growing gap between awareness and action reveals that we need to think about fundamentals—to question the underlying worldview that has led us to this critical ecological moment, to challenge our most basic cultural narratives, and to change the stories we tell.

Many people already recognize that we need to redefine and rebuild our relationship with the natural world; they just have difficulty imagining how. And that's the thing about human exceptionalism. It gives us the sense that there is no alternative or structural reimagining of our way of life. It fuels our collective inertia and disempowerment in the face of unsustainable growth and production—trends that, though relatively recent, appear inevitable. People assume these are powerful forces that we can't possibly counteract. Corporate greed and deception, capitalist economics, and a lack of political will certainly play an outsize role. But we don't just need an overhaul of these institutions; we need a new *relationship* to the world. The process of building and sustaining that relationship does not proceed entirely from the top down. It can be enacted only by individuals who are motivated by another vision and experience of what the good life might be, who can *reimagine* this richer relationship and are already bringing it into being. Though the ideology of human supremacy has shaped the dominant culture for centuries, it is rarely named or acknowledged. In this way, it remains invisible while other causes—fossil fuels, habitat destruction, biodiversity loss—are identified and studied. Much of human exceptionalism's power is drawn from this invisibility. It's the most powerful unspoken belief of our time.

The unprecedented scale of the human impact has led many scientists to name the present geological time period the Anthropocene—which recognizes *anthropos* (Greek for "human") as a major planetary force. While there may be legitimate scientific reasons for the term, it has drawn criticism from various scholars. For one, the Anthropocene discourse portrays the human impact as somehow "natural"—a species-typical behavior rather than an expression of a particular time and culture. Moreover, in assuming a role of "humankind," the so-called age of man overlooks a critical social dimension—namely that the humans least responsible for the ecological crisis are the most at risk of its consequences.

Similarly, treating human exceptionalism as some universal foundation of human thought only reinforces the colonial mindset in which this worldview is deeply enmeshed. The truth is that not all humans uphold an essential hierarchical divide between humanity and the rest of Nature. And as we'll see, we have much to learn from these alternative ideologies and relationships. Though the myth of human exceptionalism can be found in various traditions, it has been expressed and elaborated most prominently in Western culture. I routinely use language like "we" and "ours" in reference to this dominant culture in which I have spent most of my life. But again, human exceptionalism is neither culturally nor individually universal. Even throughout Western history one can always find countercultures deviating from its central dogmas in various ways.

If our titular arrogant ape isn't all of humankind, who—or what—does it represent?

In the ancient Greek tradition, drama was a way of examining what it means to be human. Protagonists often displayed hubris, an excessive pride resulting from an overestimation of their own competence and accomplishments. As I was doing research for this book, the metaphor for our current relationship to the rest of life on earth became increasingly evident. To Greek audiences, hubris was considered a "pride that blinds" because it led characters to behave in ways that belied common sense and defied the natural order, ultimately occasioning their downfall.

Consumed by a human superiority complex, the arrogant ape mirrors Hamlet—a hubristic character caught in a tragedy of its own making. *The arrogant ape is thus not a species or a culture or even an individual*—it is a way of acting and moving and being in relation to the rest of Nature. Many of us have learned the script and perform it dutifully: a role, a persona, a facade. This disguise obscures who we really are—a way to cover up our own insecurities, perhaps. Yet behind this mask is a much richer and truer way of being. When we stop making man the measure of all things, we learn so much more about other species, ourselves, and our place on this shared planet. We come to realize that human exceptionalism is the mask that disenchants the world.

Anthropocentrism likely resides somewhere in many of us. Yet as hubristic tendencies go, we remain partly or wholly unaware of it. Unlearning human exceptionalism has the power to open the door to radical transformation, to character developments within each one of us.

To some extent we are all complicit and conditioned by this worldview. But we are not determined by it. We can find other ways.

And that is why this is also a story about hope.

CHAPTER TWO

The Unlearning Curve

A S A LITTLE GIRL, I EXPERIENCED NATURE AS AN ENDLESS playground of wonders, mysteries, and adventures. The woodsy home where I grew up in Pennsylvania had a small pond in the back, which my friends and I would explore with sticks, nets, and surprised giggles. Memories of my childhood encounters with other species there remain visceral. I can still smell the damp moss after it rained and the earthy petrichor rising from the soil. I can hear the honey bees in the mess of wildflowers and vines that disguised our mailbox. Even now, I can feel the slimy string of toad eggs my dad once draped around my neck, drops of cool water trickling down my back. Perhaps because I was an only child, I found companionship in these woods relatively easily. I befriended other beings, the gray squirrels and the robins and the woolly bear caterpillars, imbuing them with rich mental worlds—with feelings and thoughts of their own. Among those I saw as siblings was my parents' Shetland sheepdog G.B. (short for Goofball), a role she herself was initially reluctant to adopt (after all, in her view, she was the firstborn).

As most parents can attest, children have a remarkable affinity for other species. The renowned sociobiologist E. O. Wilson popularized the term "biophilia" (from the root *bio*, meaning "life," and *philia*, meaning "love of") to describe this natural tendency to seek connections with other forms of life. According to the biophilia hypothesis, our attraction to other creatures is a biologically based need, one we are genetically predisposed to as a species. This hypothesis is supported by the fact that strong biophilic tendencies are present from a young age.

"I Like Them to Stay Standing Up"

Do a quick search on YouTube and you will find plenty of evidence for children's spontaneous interest in and concern for other animals. In one of my favorite videos, Luiz Antonio, a toddler in Brazil, sits behind a plate of octopus gnocchi while talking to his mom in the kitchen. Over the course of their conversation, it dawns on him that the octopus dinner in front of him was once a real live octopus. The transcript below is translated from Portuguese and edited for brevity and clarity:

> **Mom:** Now eat your octopus gnocchi.
> **Luiz:** OK. . . . This octopus isn't real, right?
> **Mom:** No.
> **Luiz:** Then alright. He doesn't speak and he doesn't have a head, right?
> **Mom:** He doesn't have a head. Those are only the chopped little legs of the octopus.

Like many kids who first notice links between what they consume and what they know about the biological world, Luiz ponders whether the octopus has any attributes of a living being. He later tries to grasp how the "legs" ended up on his plate, and what became of the octopus's head. His mom explains that it's at the fish market, where the butcher chopped it. The conversation continues as Luiz tries to resolve his confusion.

Luiz: Why?
Mom: So we can eat it. Otherwise we'd have to swallow it all.
Luiz: *But why?*
Mom: So we can eat it, love. Just like a cow is chopped, a chicken is chopped.
Luiz: Ah, the chicken . . . Nobody eats chicken.
Mom: Nobody eats chicken?
Luiz: No, those are animals!

Luiz reasons that because octopuses are animals like chickens, people do not (or should not) eat them. He then elaborates, before probing further.

Luiz: All of them are animals. Fish are animals. Octopus are animals. Chicken are animals. Cows are animals. Pigs are animals.
Mom: Yeah.
Luiz: So when we eat animals, they die?
Mom: Ah yeah.
Luiz: Why?
Mom: So we can eat, love.
Luiz: Why do they die? I don't like that they die. I like them to stay standing up.

As the clip concludes, Luiz Antonio's mother, moved to tears, asks her son to eat just the vegetables on his plate. The video today has millions of views. It's shared so widely because it shows how human-centric values—in this case, the casual consumption of other animals—belie common sense for young children. It is one of countless examples of kids extending moral consideration to other species in ways that many adults readily discount.

But what does the science have to say? Recent research in developmental psychology has put some of these ideas to the test. In one 2021 study, Yale University researchers presented American children (five to nine years old) and adults with moral dilemmas pitting varying numbers of humans against varying numbers of either dogs or pigs. The hypothetical scenarios depicted two sinking boats that contained either human or other animal passengers who could not swim, and participants had to choose one boat to save (there was also a third option to not decide). Researchers found that children were considerably less likely than adults to prioritize humans over other animals. Children often opted to save multiple dogs over one human, with many valuing the life of a dog as much as that of a human. While children generally valued pigs less, the majority still prioritized saving ten pigs over one human. In striking contrast, almost all adults chose to save one human over even one hundred dogs or pigs.

These findings indicate that human exceptionalism is not inborn, a bias intrinsic to our biological makeup as human beings. Instead, they suggest that the belief that humans are morally special is socially acquired, an aspect of our culture. Researchers interpreted their findings through the related concept of "speciesism"—that is, the tendency to grant moral priority based on species membership. They argue that children learn to prioritize humans over other animals (a.k.a. anthropocentric speciesism) only as they gain experience and knowledge of how humans use animals in our society. Most young children have little to no direct exposure to practices like meat production or animal experimentation. But as these practices become more salient—usually in adolescence, at least in Western industrialized cultures—their moral focus may become increasingly human-centric.

From childhood, our natural affinity for other species (biophilia) comes into direct conflict with these practices. We attempt to minimize the resulting dissonance in different ways. For example, children's storybooks typically portray happy farm animals living in idyllic settings instead of the intensive confinement in which most suffer, giving kids the sense that these animals lead relatively good lives. One study even found that elementary school children rated farm animals as better off than other types of animals (such as companion and wild animals). Whereas 26 percent said that farm animals sometimes feel unhappy, 46 percent said this was true for pets, and 53 percent said as much for wildlife. Most kids are unaware of the diverse ways that humans use other animals. In the second half of the study, researchers showed children pictures of common animal products (such as hamburgers, cheese, ice cream, leather jackets, wool blankets, and so forth). On average, first and third graders did not identify half of the items as coming from animals, and fifth graders missed roughly one in three. Results like these demonstrate that children frequently do not associate everyday animal products with living animals. And as Luiz Antonio's octopus revelation shows, when we do connect the dots, it usually doesn't sit right.

I recall very well the first time I saw a poultry truck roaring down the highway. There must have been hundreds of small, rusty cages stacked atop of one another, each crammed with several wide-eyed chickens, loose feathers flying out behind the truck and falling into a trail that would never find them alive. I immediately burst into tears and could not shake the vision for weeks. From that point on, my mom had to tell me to turn away every time she saw a poultry truck coming, instructing me to open my eyes only when it was out of sight.

That act of "turning away" is something we get better at over time. It's one way we cope with the difficult reality of living in a culture where the exploitation of other species is common practice. The industry knows it as much as we do, which is why there's a very deliberate disconnect between what we consume and the life that once was, at least in most modern industrialized contexts. Factory farming operations and animal laboratories are strategically located in remote, inaccessible places. Species used in agriculture and invasive research receive far less media attention than do other types of animals, with most popular educational images and films featuring wildlife instead. And because many people find it challenging to consume animal products if they resemble live animals, body parts associated with a life or personality—the eyes, face, feet—are rarely eaten or marketed. Nonetheless, in some societies, these animal parts are readily sold and consumed, a further testament to the role of sociocultural learning in shaping these perceptions.

Nothing Goes Without Saying

Another way we manage dissonance and devalue animals' lives is by denying their subjectivity—the presence of rich internal experiences. For most animals that humans put to use, there is a widespread refusal to acknowledge them as sentient individuals with meaningful emotions and thoughts. Consider how we use different names for the consumed animal and the live animal. Cows become "beef," calves become "veal," and pigs become "pork." It is noteworthy that these abstractions appear more necessary for mammals than for animals more distant from us evolutionarily, such as chickens or lobsters. Yet even when we use the same word to indicate the consumed animal and the live animal, we use a singular noun without an article. People eat "chicken" not "the chicken" or "chickens."

Researchers in cognitive linguistics have long argued that language is a system that we use to understand and relate to (and not just talk about) the world, a window into how our minds and cultures operate. The worldview of human supremacy is deeply hidden in our everyday words and phrases, attesting to its powerful grip on the public imagination.

Like terms that dissociate animal products from living animals, seemingly ordinary words legitimize human exceptionalism by describing the natural world as a commodity whose ultimate purpose is to be used and managed by humans. Sociologist Eileen Crist calls attention to renaming animals "livestock," trees "timber," rivers "freshwater," seacoasts "beachfront," mountaintops "overburden," and so on. Such terms reify beliefs that living systems matter only for their instrumental value to humans, stripping them of their inherent worth, interests, and identities. Even dictionaries, which are seen as "objective" authorities on language issues, exemplify these biases. Definitions of other animals and plants tend to emphasize their utility over key features of their biology, psychology, or behavior. For instance, ask Google to define "anchovy" and you will get the following definition from Oxford Languages: "a small shoaling fish of commercial importance as a food fish and as bait. It is strongly flavored and is usually preserved in salt and oil."

We'll speak offhandedly of the earth's "natural resources" and "ecosystem services," likely unaware that these *very expressions* define the environment around us in terms of our potential to use and possess it. In the ultimate twist, phrases like "developing" the land cast destroying forests and animal habitats for our construction projects in a positive light. We describe species used in farming and hunting as "crops" or "surplus," and lethal acts as "culling," "harvesting," or "managing" their populations. Our language obscures what really happens (killing of species and individuals) while affirming and celebrating human dominion and mastery over the natural world.

Wesleyan psychologist Scott Plous has shown how this linguistic trickery is often directly encouraged. For instance, information distributed by the 4-H club (a major U.S. youth development organization that involves teaching students agricultural literacy) warns participants in animal fairs to use terms like "chicks," "calves," and "lambs" (not "babies"); "farrow," "hatch," "foal," and "bear" (not "giving birth"); "process" (not "kill" or "slaughter"); and "health products" (not "drugs"). The idea is to avoid "humanizing" animals in a way that would draw criticism from the public. Similar word substitutions abound in scientific journals. For instance, the *Journal of Experimental Medicine* advised authors to substitute words like "intoxicant" for "poison," "fasting" for "starving," and "hemorrhaging" for "bleeding." Euphemisms like "sacrifice" and "euthanize" intentionally blunt knowledge of what is actually done after invasive experiments. Some journals explicitly counsel authors not to refer to animals by name or initials but instead to use numbers. Some even renumber (e.g., rabbit 10-8 instead of rabbit 108) to give the appearance that fewer animals were used. In this way, we are culturally conditioned to approach other species as objects rather than as subjects.

And then there are pronouns. I still remember the first time an editor crossed out "who" and replaced it with "that" every time I referred to an individual chimpanzee in one of my studies (e.g., "the chimpanzees who groomed" became "the chimpanzees that groomed"). According to most grammar authorities, "who" is always used to refer to people, whereas "that" is always used to refer to objects. We would never deem it appropriate to refer to another human being as "it"; yet, except for companion animals, this is common parlance for most other species. There are promising signs that this is changing. In 2021, a group of more than eighty experts in animal welfare, including the world-famous primatologist Jane Goodall, signed a letter calling on the editors of *The Associated Press Stylebook* to recommend that journalists use gender pronouns for other animals: "When gender is known, the standard guidance should be, *she/her/hers* and *he/him/his*, regardless of species. When it is unknown, the gender-neutral *they*, *he/she*, or *his/hers* should be used. It is also preferable to use *who* rather than *that* or *which* when describing any individual nonhuman animal."

But going one step further, consider how human exceptionalism lives on in the essence of this last phrase: "nonhuman animal." Given that the vast majority of life on earth is not human, this term is like calling a human a non-chimpanzee, or a chimpanzee a nongrasshopper! "Non-" groups millions of species together by an absence, by their failure to conform to the human archetype. Even the term "animal" itself, as the philosopher Jacques Derrida once elaborated, lumps an unthinkable diversity of beings into one homogenized category, with little in common beyond the fact that they are not human. As we'll explore later, various linguistic conventions blind us to rich cognitive and emotional worlds of other species, referring to "positive/negative states" rather than emotions like joy or fear and relegating behavior to biology and instinct rather than conscious choice. Such conventions normalize the objectification of other lifeforms and undermine their ethical standing. As botanist and Indigenous scholar Robin Wall Kimmerer puts it, "The arrogance of English is that the only way to be animate, to be worthy of respect and moral concern, is to be human."

Most of us don't think twice about these words, because their true meaning gets masked by their normality. Given the power of discourse in shaping our reality, as children become acquainted with language, beliefs in human exceptionalism become more concrete and real.

Speciesist Hierarchies

When I was growing up, my local grocery store had a small tank filled with live lobsters. Thick rubber bands around their claws, they seemed to move in slo-mo while human shoppers frantically grabbed items from the shelves around them. Before I hit puberty and my first growth spurt, I met them at eye level. Peering into the side of the tank, I stared into their beady eyes, wondering if they were OK and what they might be experiencing. Did they know where they were? Did they understand their fate? I tried to communicate with them telepathically through the glass and even hatched a plan to rescue them. I wasn't sure if lobsters could survive in backyard ponds, but it surely seemed better than this place. One day, I arrived with every intention of putting several in my backpack only to find an empty tank. Behind the counter, I saw several grocers banding and readying a new cohort of lobsters for purchase. I realized this was an ongoing system I could not easily penetrate or thwart. It made my heart sink.

Research has shown that lobsters are solitary except when mating, meaning they probably don't appreciate being kept in a tank with other lobsters (to say the least). They need their claws to catch food and defend themselves, and can be left- or right-handed—or better, -clawed!—or ambidextrous. They can live to be one hundred years old. Though we've long deluded ourselves into thinking they don't experience pain (I recall many reassurances to this effect during summer celebrations growing up), we have compelling scientific evidence that they do, which is why boiling lobsters alive is now illegal in Switzerland, Norway, and New Zealand. But today in places like the United States, you can still purchase live lobsters on Amazon and have them delivered to your doorstep fresh.

Many people would find it unthinkable to buy a live lamb or cow to kill at home, so why is it so different for lobsters? Why is it permissible in some parts of the world to consume animals (like dogs) but unthinkable in others? How can the same species (like cows) be seen as disposable in some cultures and sacred in others? To echo George Orwell's *Animal Farm*, why are all animals equal but some more equal than others?

This brings us back to the concept of speciesism, or discrimination based on species membership. As we grow up and assimilate into the culture we know, we learn about both human exceptionalism and other speciesist hierarchies. Interestingly, research indicates that children not only have less anthropocentric speciesism but also lack the more general speciesist biases of young adults and adults. Many of these biases rest on pragmatic (mostly economic and political) grounds, shaping which animals we consume, who we welcome into our homes, and even who counts as an "animal" under the law. Some studies suggest that our concern for other species depends on evolutionary proximity (e.g., we empathize more with other primates like chimpanzees than with other animals), though it remains unclear how much of that is also attributable to cultural factors (e.g., exposure to species more similar to us in the media). Through powerful processes of enculturation, we become familiar with speciesist norms. We develop strategies to manage whatever dissonance might initially arise. We bear witness to the process of dissociating—of "turning away"—in others around us, and we internalize it ourselves. Eventually, the commodification of certain species for human ends becomes somehow "normal" and "commonsense." It's difficult to unlearn anthropocentric practices, biases, and discourses because they are everywhere around us starting from an early age.

This brings us to a crucial point about human exceptionalism: this ideology is not necessarily explicitly taught; rather, we learn and internalize it by example—by observing the world and the behavior of people around us. Exposure to practices condoning the exploitation of other species is certainly one part of that process. But the worldview that humans are intrinsically more valuable than other species is perpetuated through various other social and cultural means, including aspects of our formalized education system.

Rites of Passage

One day, I walked into my seventh-grade biology classroom, where dozens of frog "specimens" awaited on trays, the stench of formaldehyde oozing from their bodies—limbs spread-eagled and pinned down to the tray, large flaps of ventral skin pulled back, their innards exposed. I recall how their organs looked metallic, almost machine-like. Next to each frog were some tools—forceps, scissors, and a scalpel. My classmates and I nervously went to our cubbyholes, grabbed kid-sized white lab coats, gloves, and safety goggles, and stationed ourselves behind a tray.

For years, we had anticipated this very moment. Throughout elementary school, older students had regaled and menaced us with stories of frog dissection day. A girl had once fainted and got sent home. A boy had been suspended for putting a frog in the headmaster's mailbox following a friend's dare. Dissecting frogs resembled a middle school rite of passage, sort of like getting your driver's license or going to prom. Even though we had ample time to mentally prepare for this exercise, a few of us didn't want to do it. At one point, my friends and I contemplated staging a walkout—a small rebellion—but ultimately decided against it for fear of receiving a bad grade. Instead, we surrounded our teacher's desk and cross-examined her about where these frogs came from. She assured us that they were already dead, that their bodies were "sacrificed" for educational purposes. No unnecessary suffering, allegedly. I've learned since that this is always what they tell you.

The dissection commenced with opening and inspecting the contents of the frog's stomach. Eager to prove themselves, some students stepped up to the plate right away and inserted their scalpels. I hesitated and observed the room around me. Classmates emitted apprehensive laughs followed by "gross!" and "yuck," but the teacher gave them praise, as the first cut was apparently the most difficult. I concentrated and put a slit in the frog's stomach lining. I felt my own stomach churn, nauseated. The teacher approached my station and said I was doing it right, so I pressed on.

Whatever emotional discomfort and squeamishness the class expressed at the outset of the exercise seemed to fade over time. As we located and dissected other organs—the liver, heart, lungs, gallbladder—"ews" and "icks" gradually turned into "oohs" and "aahs." Some of us soon discovered that our frogs were female when thousands of eggs poured out like a tiny landslide onto the tray. A vision of the tadpoles emerging in my backyard's pond each spring flashed through my mind, but I immediately shook the thought. I couldn't think that way—about this frog being a frog, a mother, a life—if I wanted to finish the dissection without too much difficulty. So I kept going. I remember feeling surprised by how the mood in the room shifted, including my own as I developed a strange, newfound sense of ease with what I was doing. I felt like I had accomplished something. I felt grown up.

Sociologists Dorian Solot and Arnold Arluke researched students and teachers during fetal pig dissections at a Rhode Island middle school. They show that students must learn to avoid ethical and emotional conflict over dissection, and how instructors facilitate this process by reassuring students into a sense of normalcy about what they are doing. The researchers contend that the educational lesson is more about this process of emotional detachment and desensitization than about animal anatomy. They liken it to an early initiation into the scientific community. "One of the skills learned from dissection is to see the animal as a scientific tool rather than a once-living being," Solot says. "The message is that detachment from caring about animals is a key skill for scientists. If you don't have it, science may not be the right field for you."

Like many students, I became a scientist because I fell in love with the living world. Science felt like a channel for biophilia that society would consider acceptable and even respectable. But even our earliest scientific education subtly reinforces the idea that Nature is a thing to be used and manipulated, so long as it serves human knowledge and progress. We learn how to cut up frogs, fetal pigs, and sheep hearts. We learn how to capture and kill insects and pin them to display boards. We learn to uproot and press plants dry and glue them to paper. We learn that learning about Nature does not necessarily mean loving Nature.

It's easy to see how certain aspects of our early education about other species and the environment normalize human supremacy. But even school curricula explicitly designed to teach children how to care for Nature may unintentionally legitimize this worldview. Environmental education researcher Helen Kopnina observed nature-based schools in the Netherlands and Canada. She conducted interviews with children (nine to eleven years old) and school supervisors in "nature education" courses and programs. In one study, she followed school gardening projects in an urban park in Amsterdam called Westerpark. Kopnina's research showed how easily teachers slip into a form of "commonsense anthropocentrism" when responding to students and talking about the living world. For instance, the students equated "weeds" with something that needs to be categorically destroyed, not as wild plants that potentially contribute to biodiversity beyond human utility. This type of environmental education inadvertently celebrates the "metaphysics of mastery," which tends to frame Nature in purely resourcist terms.

To be sure, the fact that a growing number of schools are now integrating environmental education and climate change into the curriculum at all marks a promising shift. For decades such topics were considered extracurricular (literally, *outside of* the regular curriculum) activities. Yet still so much of our traditional schooling today both physically and psychologically removes us from natural spaces. We go inside concrete buildings and interact with Nature only indirectly via screens, textbooks, and microscopes rather than through our own intuitive senses. A number of schools built in the last century were intentionally designed without many windows because windows were thought to detract from students' learning. One popular website, schoolorprison.com, even made a game out of it. Photographs of real buildings (solicited from the public) appeared on the screen and you had to guess—is this a school or a prison? As I once saw for myself, it's hard to perform above chance.

Human-Centric Histories

The view that human beings are the most important species on earth also gets reinforced through subtle (and at times not so subtle) suggestions that human experiences are somehow unrelated to, or take precedence over, the experiences of other species. Consider the definition of "history" from the latest edition of the widely read *Penguin History of the World*:

"History is the story of mankind, of what it has done, suffered or enjoyed. We all know that dogs and cats do not have histories, while human beings do. Even when historians write about a natural process beyond human control, such as the ups and downs of climate, or the spread of a disease, they do so only because it helps us to understand why men and women have lived (and died) in some ways rather than others."

Upon first reading this, all I could think was, no wonder history was my least favorite subject in school! It centered on exclusively human actors, as did most of my other classes: literature, philosophy, art, and so forth. We call it the *human*ities, after all.

Ironically, while I never had a knack for history, I was immediately captivated by evolution—which is just history on a much longer timescale. But it's also a history that allows other species to play a part. *This* is why I was drawn to science in the first place. Aside from an hour a day spent outside for recess, science classes were the only stable part of the curriculum where nonhuman actors made any reliable appearance. I loved learning about microorganisms like amoebas and how caterpillars became butterflies through metamorphosis. I delighted in the opportunity to hear about plant photosynthesis and to funnel-feed a colorfully decorated lizard my classmates and I named Art Gecko. My interest in science and especially evolution solidified in high school biology classes when I first read Charles Darwin's revolutionary *On the Origin of Species*.

Learning about Darwin's theory of evolution by natural selection brought Nature alive in profound ways for me—ways, as a teenager, I was starting to forget as I assimilated into a culture shaped by human exceptionalism. The realization that all species are descended from common origins and bound together in the complex web of life filled me with a profound sense of wonder and gratitude. It still does to this day whenever I think about it—that I even get to be part of this sacred experiment. Where finches and fireflies, sea sponges and sequoias, hyenas and humans *all* have histories intertwined in deep time. Where even the "species" category itself is much blurrier than we conventionally assume. According to Darwin, shared ancestry on a shared planet gives rise to differences between species that are a matter of degree and not kind, a continuity not just in physical form but in psychological makeup, in the very constitution of our minds. In much of his work, including *On the Origin of Species*, Darwin does not describe other species as passive vehicles of their genes and environments. He imbues them with rich inner lives—even creatures like earthworms. Imagine! He used a language that resonated with me. He gave words and theories to my childhood experiences of kinship and respect with the living world, backing them with scientific credibility.

It is astounding how many of Darwin's principles have stood the test of time. Today, evolution is considered the most important principle in all of biology. My college biology class introduced me to the wise words of geneticist Theodosius Dobzhansky, who once famously wrote, "Nothing in biology makes sense except in the light of evolution." And yet, Dobzhansky is also known for his (rather less wise) claim that "all species are unique, but the human is uniquest."

How can someone be an outspoken proponent of evolutionary theory *and* the basic idea of human exceptionalism? Many of my students at Harvard admit to a similar paradox—choosing to concentrate in human evolutionary biology because they are drawn to evolutionary thinking *and* maintain some unshakable sense that humans are especially special. This illustrates how anthropocentric "common sense" is such a taken-for-granted view and way of life that even the staunchest supporters of evolution unknowingly succumb to it.

There is no single, linear narrative of evolutionary progress. Instead, evolution is a process of continuous branching and diversification—more like a tree or web than a ladder of life. Stephen Jay Gould, who taught right across the street in Harvard's Department of Organismic and Evolutionary Biology, vehemently cautioned against equating evolutionary change with directionality and progress. Yet the process is often interpreted or presented as progressive (or "orthogenetic"), culminating in us. Classic depictions of major evolutionary transitions will show microbes, plants, and invertebrates leading up to smaller vertebrates like fish and birds and, finally, large mammals like humans at the top. Such renderings reinforce the widely held but erroneous view that evolution proceeds from "primitive" to more "advanced" organisms. And it's also worth noting that we're talking about how this human-centric idea spreads in schools that even teach evolution in the first place!

Darwin himself maintained that "it is absurd to talk of one animal being higher than another," yet talk of "higher" and "lower" organisms is still common today. We tend to assume that our species is the end point of evolution, that there is an inexorable trend toward "human-ness." One iconic image shows the silhouettes of humans' ancestors gradually transitioning from hunched-over ape-like forms to upright cavemen-looking figures that eventuate in *Homo sapiens* (a wittier version depicts modern humans at the last stage crouched over again at their computers). People often misinterpret this image as evidence of a linear process by which modern humans descended from chimpanzees. Humans did not "evolve from" apes or monkeys. We share a common ancestor with both chimpanzees and bonobos that lived around five to seven million years ago. Chimpanzees and bonobos (and all other species with whom we share ancestry and who have not yet gone extinct) have continued to evolve in parallel as the evolutionary tree continues to branch outward. Yet linear representations tend to erase those other-than-human histories. They elide the diversification of species from common ancestral populations into one human-centric story.

To make this point clear, consider the following question: Are frogs more closely related to fish or to humans? Many people assume a closer relatedness between frogs and fish than between frogs and humans. But because frogs and humans share a more recent common ancestor (one that is not shared with fish)—frogs are more closely related to humans.

Perhaps even more surprisingly, some fish are more closely related to us than they are to other fish. For instance, lungfish are more closely related to mammals like humans than they are to other aquatic species like salmon. Evolutionarily speaking, "fish" is an invalid functional category that does not reflect real evolutionary relationships. The word aggregates tens of thousands of different species—more in number and diversity than all terrestrial vertebrates combined. As my friend the aquatic animal welfare scientist Becca Franks notes, "The vast majority of modern fish species evolved hundreds of millions of years after humans shared a common ancestor with them—which means that modern fish are not primitive fore-runners to our lineage any more than humans are primitive forerun-ners to theirs."

There is a great fallacy in making humans the reference point for the rest of the biological world. Not only because it is scientifically invalid, but because it deprives us of a richer, fuller perspective and way of life. Evolutionary biologist Robert O'Hara said it best: "When we come to realize that even among the vertebrates there are 50,000 different 'vertebrate stories,' each one with a different ending and each one with a different narrative landscape; when we truly think in terms of the diverging tree, instead of the line; when we understand that it is absurd to talk of one animal being higher than another; only then will we see the full grandeur of the historical view of life."

Folk Biological Knowledge

What do young children use as their reference point when making sense of the biological world? How do they acquire fundamental bio-logical concepts including animal, plant, and living being? Folk biology—the study of how people conceptualize and reason about nature—has generated fascinating insights into these questions.

For decades, developmental psychologists maintained that chil-dren reason about the natural world from a human-centric vantage point. This was largely based on psychologist Susan Carey's influen-tial model of conceptual development. Carey argued that children's earliest understanding of other living beings is mainly in reference to, or by analogy to, human beings. In other words, young children view humans as the prototypical animal and must undergo a major con-ceptual shift to move from this anthropocentric view to one in which they see humans as one animal among many.

One powerful form of evidence came from inductive inference tasks. In these tasks, experimenters present participants with pictures of a base item (such as a human, dog, or bee) and teach them an artificial new property about it. For example, the interviewer might show a picture of a dog and say, "Have you ever heard of an omentum? Omenta are green, round things. Here is a picture of something that has an omentum inside it." Next, the researcher tests whether the participant generalizes (or "projects") that property onto other entities (including other humans, dogs, bees, aardvarks, flowers, and so forth). Carey found that young children generalized from humans to animals based on biological similarity to humans (for example, to dogs more than to bees), but hesitated to project from animals to other animals, including to humans. These children even favored inferences from humans to insects over inferences from bees to other insects. In short, it was as if humans were the only proper base for generalization. This pattern was interpreted as evidence that young children's conception of the biological world is human-centric.

Yet as subsequent research pointed out, Carey's work was conducted almost exclusively with individuals from North American urban populations. These groups are also known as WEIRD—Western, educated, industrialized, rich, and democratic. Despite being highly unrepresentative of humanity on the whole, WEIRD people are the most common subjects of psychology research. Thus, it is not clear from Carey's original work which aspects of children's folk biological knowledge are human universals and which might rely on cultural conceptions and experiences.

To evaluate the role of cultural milieu in children's folk biological knowledge, anthropologists performed an experiment with three populations: urban Boston children, rural Wisconsin majority-culture children, and rural Native American (Menominee) children. They gave participants a property-projection task based on Carey's original paradigm. Their findings were striking: only urban children showed evidence for early anthropocentrism. Unlike their city-dwelling counterparts, rural children generalized in terms of biological similarity.

Moreover, young Native American children showed reasoning based on ecological context and relationships among species (for example, they might justify generalizing from bees to bears because a bear might acquire the property by eating the bee's honey). Researchers concluded that a lack of intimate contact with other animals and plants is responsible for the anthropocentric bias observed in urban American children. In other words, human-centered reasoning patterns reflect a lack of experience with the natural world rather than a radically different construal of it.

In rural environments, exposure to and interest in the natural world are everyday experiences. On the contrary, the only real animal that most urban people regularly encounter and know much about is *Homo sapiens*. One rather troubling statistic holds that the average American kid can recognize one thousand corporate logos but can't name ten plants or animals native to the region in which they live. This lack of diverse input likely underpins the fact that basic biological categories for WEIRD adults (and thus those learned by children) are broad lifeform categories (e.g., fish, bird, mammal, tree). For example, if one asks "What's that?" while pointing to a trout, their typical answer is "fish." However, in societies with more familiarity with Nature, the generic species (e.g., trout, crow, fox, or maple tree) forms the basic category.

Researchers in Japan provided further support for the idea that anthropocentrism is heavily mediated by children's experience. One study compared urban Japanese children who had cared for and raised goldfish in their homes with those who had not. When researchers assessed these children's attributions of biological properties, they found that children without goldfish-raising experience showed an anthropocentric pattern but that those who had raised goldfish did not.

Despite clear evidence that experience matters, these intriguing patterns left one important question unanswered. Namely, might all young children begin reasoning from an anthropocentric stance, but those with relatively enriched experience with other species leave this anthropocentric perspective behind sooner than do those with rather limited contact? To test this, Patricia Herrmann and colleagues at Northwestern University developed a modified version of Carey's induction task that allowed them to test the inductive reasoning of urban participants as young as three years of age. They demonstrated that anthropocentrism is not the first developmental step even in urban children's reasoning about the biological world. Although urban five-year-olds adopt an anthropocentric perspective, replicating previous reports, three-year-olds show no signs of anthropocentrism. This suggests that anthropocentrism is not an inherent part of early conceptual development, but is instead an acquired perspective that emerges between the ages of three and five in children raised in urban environments.

What evidence is there that folk ecological knowledge affects how we treat the natural world in practice? One case study involved three cultures who depend on the same habitat in Guatemala's northern rainforest: Itza' (Native lowland Maya), Q'eqchi' (immigrant Maya from neighboring highlands), and immigrant Spanish-speaking Ladinos (mixed Amerindian and European descent). Researchers measured plot sizes, species diversity, tree counts, canopy coverage, and consequences for soils in these regions. Controlling for important factors like age and income, they found that a lack of knowledge about the natural world predicts unsustainable agroforestry. Q'eqchi' immigrants' folk ecology conceptualizes plants as resources to be exploited (plants are passive donors to animals, and animals have no effect on plants), and their agricultural practices are correspondingly most insensitive to forest survival. On the other hand, Native Itza' folk ecological models stress reciprocity in animal-plant interactions (animals can either help or harm plants), and their practices respect and preserve the forest. Immigrant Ladinos display nonreciprocal folk ecology (where plants help animals but animals do not help plants), and their agroforestry patterns are intermediate. Such studies are important because they demonstrate that sustainable practices are more likely in communities where humans and other animals are seen as participants in an intricately connected circle of living beings. They demonstrate that mental models of Nature (what we know about Nature and how we see ourselves in relation to Nature) predict how we treat Nature (and what we do about Nature) in practice.

Ecological Amnesia

So many aspects of our urbanized lifestyles limit contact with and attention to the natural world. Author Richard Louv coined the phrase "nature-deficit disorder" to describe how humans—especially urban children—are spending less and less time outdoors. "The child in nature is an endangered species," Louv contends, linking this deprivation to attention and behavioral difficulties, diminished use of the senses, and higher rates of emotional and physical illnesses. What's more, opportunities to interact with Nature are harder to come by as the environment continues to degrade, but the trend may occur outside of our awareness, as a kind of amnesia.

Marine biologist Daniel Pauly's "shifting baseline syndrome" concept was the first to describe how cultures become progressively blind to environmental degradation. Today, people also know it through the related concept of "environmental generational amnesia." The idea is simple: all of us develop an idea of what is environmentally "normal" based on the natural world we encounter in childhood. With each successive generation, environmental degradation generally worsens, but each generation takes that deteriorated state as the new norm. What we see as degraded today our children will see as "natural," and what we see as pristine Nature today our ancestors would see as highly degraded.

One telling example of this phenomenon comes from the Yangtze River basin in China. A river dolphin, also known as the baiji, was declared extinct in 2007 because of human pollution, noise, and overfishing. Researchers found that this large, charismatic species was rapidly forgotten about by local people soon after they stopped encountering them on a fairly regular basis. One of the key predictors was youth. The younger the participants, the less likely they were to know about the baiji, despite being prompted by photos, local names, and descriptions.

The point here is that our relationship to Nature is not merely about socialization and cultural influence. It's about how the degraded natural world leads kids *themselves* to construct mental models of Nature in which the situation is effectively "normal" when it's not. And by extension, how a human supremacist culture gets further and further normalized, until it's hardly even visible or noteworthy. Of course, children's experiences with the natural world are varied, including not only their habitual surroundings (e.g., rural versus urban), cultural models and narratives, socioeconomic status, and formal educational training, but also informal learning opportunities afforded by the mainstream media, companion species, and leisure activities like camping trips and visits to the zoo.

At zoos, I often find myself more interested in observing the human animals than the animals most people buy tickets to see. I am always struck by how indifferent young children seem toward the zoo's "headliner" species like polar bears and elephants. They are not so excited watching a lone Siberian tiger pace back and forth through his enclosure or being bystanders to a sea lion feeding in the distance. They appear much more interested in getting up close and personal with goats at the petting zoo or witnessing conversations between terns in the aviary. The adults are often less curious and appear to see these behaviors and interactions as relatively mundane. One of my colleagues reached a similar conclusion in a study he conducted at Royal Burgers' Zoo in the Netherlands, which placed undergraduate student observers among the public. Observations of the chimpanzee exhibit revealed how children took everything that the chimpanzees did seriously and became concerned upon viewing aggressive behavior. The adults, on the other hand, waved it all off as fun and games and didn't bother to understand what the apes were doing. Whereas children watched in fascination and were more accurate in their interpretation, adults mocked and trivialized the scene before their eyes. It's as if in adulthood we forget how to really *see* other species in ways that are second nature when we're children.

To appreciate the central role that other species play in children's lives, one need look no further than their books, movies, and other media. On the surface, this may seem like another piece of evidence for their biophilia. But it also signifies something deeper. As novelist John Berger wrote in 1980:

"Children in the industrialized world are surrounded by animal imagery: toys, cartoons, pictures, decorations of every sort. No other source of imagery can begin to compete with that of animals. The apparently spontaneous interest that children have in animals might lead one to suppose that this has always been the case. Certainly some of the earliest toys (when toys were unknown to the vast majority of the population) were animal. Equally, children's games, all over the world, include real or pretend animals. Yet it was not until the 19th century that reproductions of animals became a regular part of the decor of middle class childhoods—and then, in this century, with the advent of vast display and selling systems like Disney's—of all childhoods."

Even the iconic teddy bear did not appear until the early twentieth century, following a hunting trip attended by President Theodore Roosevelt. Roosevelt's attendants decided to chase, club, and tie a small black bear to a tree and invited the president to shoot her. As the story goes, he declined because it would be unsportsmanlike (but still ordered the bear to be killed and put out of her misery), which inspired the advent of the beloved toy.

As this example reveals, the great popularity of animal imagery in children's lives may parallel our separation from and supposed superiority to the rest of Nature. It's as if these other representations are compensating for a relationship we have lost and are struggling to remember.

———

A 2014 exhibit by artist Jenny Kendler at Expo Chicago called *Tell It to the Birds* featured an interactive sculpture inviting participants to share a secret with the natural world. Visitors would enter an organic dome-like structure resembling a bird's nest or animal shelter. Inside was a lichen "dish" concealing a microphone, with custom software that live translated the secret into one of eleven birdsongs belonging to a species under threat. Though the prompt was rather broad, the implicit suggestion was that of a confessional, an opportunity to disclose a sense of wrongdoing or nostalgia when it comes to our relationship with Nature. It became a place for people to share a childlike sense of affinity with other species that had since been neglected or forgotten, and a desire to return to it. As Kendler explains:

"It is common to hear that many of us felt closer to nature when we were children. Literally closer to the earth, we crouched downward to examine beetles, pick up feathers and to smell dense piles of moss. We picked dandelions that grew up through the pavement. We were not concerned about getting dirt under our nails. As adults, however, many of us find it is no longer 'sensible' to investigate the earth. We no longer speak back to birds. Many of us have forgotten our intimate connection to our former world, having been tasked by 'growing-up'—moving from the wild ways of childhood into the studied and arranged culture of the adult."

Children are born assuming an intimate connection with other living beings. This biophilic sense is part of not just our developmental history but our evolutionary heritage as a species. It's easy to overlook in the human-dominated landscapes in which most of us live today that our sense of superiority to the rest of the living world is highly atypical from the long perspective of human evolutionary history. And thus, it is only natural for people to feel dissonance and remorse about harming other species and the environments we cohabitate.

So we turn away. We try to disengage. We may take comfort in the platitudes we heard growing up. We see other good people around us doing the same, and we all just hope and believe that one day it might be different. We resign ourselves to the idea that it's "just how it is" for now. What we don't often realize is that in doing so, we legitimize human exceptionalism even further. We normalize the sense that we are somehow distinct from and superior to Nature with entitlement over her "resources," a narrative with dire consequences we can no longer punt for future generations to resolve.

But more than that, the human-centric worldview deprives us of richer and more meaningful experiences in the *here-and-now*. Of childlike attention and scrutiny of tiny creatures. Of feeling connected to something greater than the self. You experience it on a cellular level. It's what kids feel when they mourn cars running over earthworms migrating across the pavement. Or when they realize that the food they just consumed was once home to a life. They are socialized to shut off from these difficult realities. But there's the rub: it's this same channel that affords a sense of awe the first time they see fireflies light up the night or witness the sky darken as flocks of birds migrate south for winter. It's the same one that leads them to forge empathic relationships with the diverse beings who inhabit their houses and backyards. As much as we may try, we can't simply block this channel in one direction. As kids, we see Nature as animated and alive. Can we return to that sense of wonder, curiosity, and rapt attention? Can we view ourselves as animals, mammals, and primates after all? Human exceptionalism may sometimes seem inescapable, but it is not. It is not a cultural universal, nor is it the lens we come into the world wearing. Historically speaking, it has been a part of our society only in the mere blink of evolutionary time called modern civilization. But because it is socially learned, it can also be unlearned. I liken it to an unlearning curve. Because once you're on it, once the veil of human exceptionalism is lifted, you start to see it everywhere.

CHAPTER THREE

A Compulsion to Contrast

I F HUMAN EXCEPTIONALISM IS THE PRODUCT OF THE DOMI-
nant culture, not something intrinsic to human nature, then how
did this worldview come about, and ultimately gain such a powerful
grip on the public imagination? From Aristotle to the Bible, from the
Enlightenment to Marx, the idea of the human as a superior animal,
or a being that is different from an animal altogether, permeates
Western thought.

There has been no shortage of qualities put forward to justify this
claim. Rationality, language, tool use, culture, innovation, morality,
consciousness—the list goes on as we've tried continuously to distinguish ourselves from the animal kind. Political scientist John Rodman called this the "Differential Imperative" because it resembles
our almost compulsive need to identify differences between us and
other species that confirm our supposed superiority.

The Differential Imperative is often said to have co-emerged with
the rise of human civilization and especially plant and animal agriculture. The critical period that took *Homo sapiens* from nomadic
bands of hunter-gatherers to sedentary farming villages—also known

as the Neolithic or Agricultural Revolution—began some ten thousand years ago. Forests and grasslands were cleared to make way for crops. Water-control systems diverted rivers into canals and impoundments for irrigation. The domestication of various plants and animals created food surpluses, spurring population growth and allowing people to consolidate in larger urban centers. This transition to settled farming radically changed our relationship with the natural world. Increasingly, we lived in environments of our own making, a human realm set apart, a "culture" cordoned off from "nature." Of course, historians debate how revolutionary this all was. Reality is often much more complicated, and early human societies were far more diverse than is commonly believed, as were their relationships to land and agriculture. I don't dare tell the history of humankind in a single chapter, or book for that matter!

Suffice it to say that when humans distanced themselves from and exerted dominance over Nature, the repercussions were not merely practical but also profoundly psychological. By the time we reach classical antiquity and the ancient worlds of Greece and Rome, the human mind in these societies was already preconditioned favorably toward the Differential Imperative. Sociologist Eileen Crist points out that anthropocentrism likely solidified incrementally over time through an *interplay* between a newfound capacity to control and manipulate Nature, on the one hand, and ideas about human distinction that gained cultural traction, on the other.

The Great Chain of Being

One of those pivotal ideas is often traced back to the Greek scientist and philosopher Aristotle. In his quest to understand order and relationships in the biological world, Aristotle devised a classification scheme known as the *scala naturae* (*scala* translates to "ladder" and *naturae* to "of nature"). As its name suggests, the *scala naturae* arranged all matter and life on earth into a hierarchical structure. Human beings occupied the top rung of the ladder, with other animal forms—starting with mammals like us and descending through reptiles, amphibians, fish, insects, mollusks, and so forth—occupying lower rungs. Still lower were plants and fungi, followed by minerals and other inorganic objects on the bottom rungs.

Aristotle based his classification system on various characteristics he could observe in the natural world. For instance, he maintained that all animals and plants can grow and reproduce, making them superior to minerals. Yet apparently only animals can move and perceive, giving them primacy over plants. Humans alone, according to Aristotle, were endowed with the capacity to think rationally, placing us at the pinnacle of the hierarchy. As scholars like Gary Steiner have argued, once philosophers began to assert that reason distinguished human beings from "the beasts," the moral status of other species shifted fundamentally. Speciesism would really take hold with the Greek and Roman Stoics, but the notion was already incipient in Aristotle's *Politics*:

"After the birth of animals, plants exist for their sake, and . . . the other animals exist for the sake of man, the tame for use and food, the wild, if not all, at least the greater part of them, for food, and for the provision of clothing and various instruments. Now if nature makes nothing incomplete, and nothing in vain, the inference must be that she has made all animals for the sake of man."

Aristotle's *scala naturae* was the greatest biological synthesis of its time, and his work remained the secular authority on Nature for centuries to come. Aristotle was clearly not an evolutionist but rather an essentialist; for him, species have never changed. All were created in their present, immutable forms with an assigned place in the natural order. The idea of a fixed scale moving from least to most complex and perfect, is, in many ways, still with us today. It survives in a number of guises—even, as you may recall, in evolutionary depictions.

Just the other day I came across an article in which researchers analyzed papers published between 2005 and 2010 in top scientific journals. They found that progressionist linear language and concepts (so-called ladder thinking) is still present. Even the most specialized journals in evolutionary biology still reference "lower" and "higher" lifeforms. If they are not free from the problems of *scala naturae* language and thinking, what about the nonexperts?

In late Christian Rome and into the Middle Ages, particularly during the rise of medieval Scholasticism, Aristotle's ladder extended beyond the earthly realm into the heavenly spheres. It was eventually incorporated into the idea of the Great Chain of Being, which places God at its summit followed by angels and spirits, humans, other animal and plant life, and ultimately lifeless matter at the base. Humans alone were thought to possess both divine powers such as reason and imagination (like those ranking above them) and material bodies (like those below them). They therefore straddled the world of spiritual beings and the world of physical creation, standing closest to the perfection, omniscience, and omnipotence of the gods. The Great Chain of Being influenced philosophers, scientists, and theologians for years to come. Anticipating the biblical story of human dominion, the ancient Greek playwright Sophocles celebrates man's exceptional powers over the natural world—land, seas, winds, wild and domesticated animals, and more. "Many the wonders," rings the chorus of his famous *Antigone*, "but nothing more wondrous than man."

"Let Them Have Dominion"

In the classic 1951 film *The African Queen*, Humphrey Bogart defends his prior drunken evening by appealing to "human nature," prompting Katharine Hepburn's famous reply: "Nature," she says as she momentarily looks up from her Bible, "is what we are put in this world to rise above."

Is it a coincidence that she is holding the Bible as she delivers those memorable lines? Likely not. Many scholars attribute the dissemination of anthropocentric beliefs, and our corresponding sense of superiority to Nature, to the Judeo-Christian religious tradition. Scripture and the writings of Christian thinkers like Saint Augustine and Thomas Aquinas are often noted for their remarkably anthropocentric character.

The text that best exemplifies this view comes from the book of Genesis. From the Bible's opening words, we are told that what sets humans apart from other species is that we alone are created in the image of God. This privileged position grants us ownership over other living beings:

"And God said to them, 'Be fruitful and multiply and fill the earth and subdue it, and have dominion over the fish of the sea and over the birds of the heavens and over every living thing that moves upon the earth.'"

This passage has been interpreted as condoning an instrumental view of Nature in which other species exist for the sake of humankind. Historian Lynn White Jr. wrote undoubtedly the most controversial and widely cited paper on the topic. In his article, published more than fifty years ago in *Science*, White famously argues that the ideological source of the Western world's environmental problems was "the orthodox Christian arrogance toward nature." In his view, the most powerful religious doctrine of all established a chasm between man and Nature and insisted that it was God's will for humans to exploit the natural world.

Among White's crucial insights concerns the influence of Christian anthropocentrism in shaping contemporary "secular" beliefs and institutions. He provocatively argues that Christian dogmas—including a belief in humanity's rightful mastery over Nature and an implicit faith in perpetual progress—pervade ideologies ranging from Marxism to scientific rationalism. With a linear conception of time, the biblical creation story departs dramatically from cyclical views of time espoused by Aristotle and other intellectuals of the ancient West. And this idea of linear progress feeds directly into beliefs in human ascendancy. As White describes:

"By gradual stages a loving and all-powerful God had created light and darkness, the heavenly bodies, the earth and all its plants, animals, birds, and fishes. Finally, God had created Adam and, as an afterthought, Eve to keep man from being lonely. Man named all the animals, thus establishing his dominance over them. God planned all of this explicitly for man's benefit and rule: no item in the physical creation had any purpose save to serve man's purposes. And, although man's body is made of clay, he is not simply part of nature: he is made in God's image."

I wouldn't recommend getting baptized as a teen. To my family's total disbelief (and mine still today), I insisted on receiving a formal church baptism at the peculiar age of thirteen. This was an especially strange request because neither of my parents was particularly religious—my mom was scarred from strict Catholic schooling growing up, and my dad was a devout atheist who preferred to avoid church altogether. But my grandmother sang alto and played the drums at her local Presbyterian church, which perhaps warmed me to the idea.

Not one to voluntarily get dolled up or out of bed before noon, I threw on a honeydew-colored dress just in time for the Sunday morning service. There were two other baptisms before mine, but we were all called up to the altar area simultaneously. I was the only one wearing green and walking on two feet; the other two were infants, draped in white and wriggling about in their parents' arms. The reverend intoned a short prayer and slowly raised the water bowl from the table. He went down the line, the little ones before me looking perplexed as he splashed water over their heads. I closed my eyes in anticipation as he motioned toward me. The reverend was rather short and stocky, whereas I was a tall, gangly teen, meaning he had to stand on his tiptoes, barely getting the water to reach my head. This caused members of the audience—including my own father!—to suppress (and ultimately not suppress) laughter. The whole affair was embarrassing and less sacred and sanctimonious than I'd hoped. I received no gifts, no party—perhaps my ulterior motives from the start.

But I've often wondered whether my motivation stemmed from something much deeper. I would describe my childhood as a spiritual (though not necessarily formalized religious) one. Many people would say as much. It was inherent in my attempts to commune with the animals and plants, in my regular contemplations of Nature's mysteries and majesties, and my place in all of it. Yet alongside increasing exposure to a culture of human exceptionalism, adolescence was a time of human-centered growth. My human friendships, romances, grades, and material possessions, the latest trends—these things mattered. Perhaps as I sensed being pulled away from the grander community of life, I sought communion and deeper meaning elsewhere. At the same time, I never entirely shook my animistic sense that life and agency were everywhere around me. Somewhere along the line, I had heard that people were condemned and punished for eternity for even thinking stuff like this. Who wants to risk ending up in hell?

The orthodox Christian doctrine damned pagan nature worship. According to Lynn White Jr.'s thesis, the emergence of Christianity marked the moment humans broke away from ancient beliefs that all beings, including other animals and plants, were imbued with spirits or souls. Over time, so-called nature religions that valorized nonhuman deities were gradually replaced by religions venerating humanized gods. And this had profound psychological consequences. "The victory of Christianity over paganism," White remarks, "was the greatest psychic revolution in the history of our culture." His words have become pivotal talking points for the modern environmental movement.

To be sure, human subjugation of the natural world is far more complex than White made it out to be, as is the Christian faith itself. Some historians point to new commercial incentives, rather than the replacement of pagan animism by Christianity, as fueling the large-scale destruction of Nature. Karl Marx, for instance, argued that the emergence of private property and a monetary economy led Christians to exploit the natural world; according to him, it was "the great civilising influence of capital" that finally curtailed the "deification of nature." More and more scholars today claim that Christianity is actually ambiguous about the man-nature relationship, or even consistent with an environmental ethic. There are alternative interpretations of biblical phrases like "have dominion over" and "subdue the earth" with a somewhat less anthropocentric ring to them. There are key historical figures who didn't seem to subscribe wholesale to a natural hierarchy, including Saint Francis, who famously tried to set up a democracy of all God's creatures. Legend has it that Francis preached to birds. In a story told in his biographies and commemorated in a beautiful fresco by Giotto, a flock of birds surrounded him as he delivered an impromptu sermon to them. He rejoiced at the birds and wondered aloud to his companions why he had never preached to them before. From that day on, he made a habit of it, and throughout his life there are many remarkable events of Francis speaking with other animals and treating them as rational creatures.

Fortunately, it's not necessary to assert that Christianity is inherently anthropocentric in order to recognize that the dominant reading and reception of it has been. For every dissenter today urging a more ecofriendly reframing of Christianity, there appear to be others doubling down on human exceptionalism in its name. The *Catechism of the Catholic Church*, for instance, teaches that "animals, like plants and inanimate beings, are by nature destined for the common good of past, present, and future humanity." The International Committee of Human Dignity, an organization aiming to reassert traditional scriptural values, maintains that "the true nature of man is that he is not an animal, but a human being made in the image and likeness of God, his creator." Though on the decline for years, Young Earth creationism—which holds that humans live in a God-made world whose only history is the dominance of our species—is still remarkably prevalent. According to a 2019 Gallup poll of U.S. public attitudes, when asked "Which of the following statements comes closest to your views on the origin and development of human beings?" 40 percent of respondents selected: "God created human beings pretty much in their present form at one time within the last 10,000 years or so."

Of course, if one goes searching, traces of anthropocentrism can be found in many major religions—not just Christianity. As just one example, many Buddhist teachings include the concept of a gradation of beings (as in hierarchies of reincarnation), with humans as the ultimate state. However, it is worth noting that Buddhists do not necessarily regard these hierarchies as a morally relevant distinction justifying animals' exploitation. Nonetheless, the primary reason these religions aren't explored further here is that they did not suffuse Western institutions like science and economics—and thus the adoption and application of beliefs in human exceptionalism in the dominant global culture—as profoundly as Christianity did.

Animal Automata

By the time I started university, I was confused about humans' and my own place in the cosmos. I was contemplating going to medical school, because that's what most science- and biology-inclined students around me did. But my first undergraduate biology class had me up late at night memorizing amino acid chains and DNA sequences. I wasn't so enthusiastic because—as I only later realized—I wanted to study whole organisms (a word which itself hasn't managed to escape reducing them to their constituent parts). As a junior, I was lucky enough to come across a class that would only further convince me of this.

I still remember in vivid detail the photograph my professor, the renowned primatologist Frans de Waal, projected on the first day of a primate behavior class—an image of two chimpanzees (adult males named Nikkie and Luit) up in the trees. Nikkie was displaying pilo-erection, the bristling of hairs due to anxiety or excitement. His hand was extended, palm up, toward Luit, seeming to beckon him closer. De Waal went on to explain that the photo had been taken minutes after an aggressive conflict between the two, causing them to retreat to separate branches. Yet moments after the image was captured, the former opponents came together again in a friendly embrace—a behavior known as reconciliation. I was amazed that such a behavior existed in the animal kingdom and looked so similar to our own conflict-resolution habits. I immediately became curious about whether the underlying motivations for the behavior were also shared. Coincidentally (or perhaps not at all), I ended up writing my doctoral dissertation on this topic some years later.

With my best interests in mind, an undergraduate academic adviser originally discouraged me from embarking down this career track. It was becoming harder to get funds for animal research, in part because the field of psychology (where the animal behavior courses at my university were taught) was moving toward the neurosciences. But much more than their brains, I wanted to understand their *minds*. I was interested in their thoughts and feelings and motives, especially as they related to social interactions. I wanted to understand what those chimpanzees were experiencing.

Merging the Christian concept of an eternal soul with secular notions of the rational mind, another intellectual tradition emerged around the Enlightenment that denied other species the capacity for any kind of experience altogether. The French philosopher René Descartes developed what may be the most explicit argument for an essential difference between humans and other animals. Like many before him, Descartes maintained that humans alone have language and thereby reason. But he added that other animals lack consciousness because, unlike humans, God did not grant them immaterial souls. And because they lack reason and consciousness, they are incapable of feeling anything. In his account, other animals are machines, or automata; they "perceive" no more than the way that thermometers "perceive" changes in temperature. Their reactions are merely mechanical responses to external stimuli. For Descartes, the human body was also an automaton, but we alone had an incorporeal (mental) existence. Dislocating the mind from its material embodiment, he developed what's now known as Cartesian (or mind-body) dualism.

My students are often baffled by the boldness of Descartes's claims that other animals have, in his words, "no experience of any kind." It seems to fly in the face of common sense and knowledge, especially among those of us who've lived with companion animals. But beyond a clear precedent in the works of Aristotle, the Stoics, and Christian theology, it's not surprising that Cartesianism gained favor at the time. It was an era growing accustomed to mechanical inventions like clocks, watches, and automata of every sort, rendering the human mind receptive to this mechanistic view of life.

Moreover, that other animals were unfeeling machines was a convenient belief in an era when vivisections—live surgeries—were becoming common scientific practice. As our capacity to probe Nature with tools and technology grew, so too did beliefs that human beings were the source of all knowledge and value (a view broadly known as humanism). Human command of the natural world was vital to the ongoing tale of humanist progress. More and more, we equated knowledge of Nature with our ability to control natural systems. Descartes himself stated this plainly: "I perceived it to be possible to arrive at knowledge highly useful in life . . . and thus render ourselves the lords and possessors of nature." It's a vision that was shared by many influential thinkers of the time. Francis Bacon, the father of empiricism who would lay the foundation for the scientific method, was particularly famous for his pontifications in this regard: "The mechanical inventions of recent years do not merely exert a gentle guidance over Nature's courses, they have the power to conquer and subdue her, to shake her to her foundations." As these forebearers of modern science and philosophy show, the notion of human distinction and the exploitation of the natural world go hand in hand.

Despite Darwin

Several years after dropping out of medical school, Charles Darwin had a once-in-a-lifetime opportunity. He was invited by Captain Robert FitzRoy to circumnavigate the globe aboard the HMS *Beagle* as the ship's naturalist. They set sail from England in 1831 and arrived four years later on the Galápagos Islands—an archipelago six hundred miles off the coast of Ecuador that straddles the equator in the Pacific Ocean. Darwin's time and observations there would go on to inform and inspire his theory of evolution by natural selection. The Galápagos animals were in ways notably different from those on the mainland as well as from island to island. These differences were early hints that species change and that this occurs in relation to their environments. Far from being immutable, life seemed adapted to each island, eventually producing what appeared to be new species.

Having been lucky enough to visit the Galápagos myself, I can only imagine what these men thought when they first landed on its shores. This rugged landscape of sunbaked lava is accentuated by craters, jagged sea cliffs, and looming volcanic mountains (many of which remain active today). Captain FitzRoy described it as "a shore fit for Pandemonium," and I can certainly see why. Yet this same unforgiving terrain shelters a fantastic diversity of plant and animal species, many endemic—found nowhere else on earth. The lowlands are home to spiny cacti and thorny shrubs of all kinds. The highlands host a variety of trees, fleeced in colorful mosses and lichens. But the islands' most famous inhabitants include its giant tortoises, blue-footed boobies, Galápagos penguins, sea lions, and marine iguanas. Darwin soon realized how tame the animals were, unaccustomed to human presence or predation. He once wrote that he nudged a curious hawk off a branch with the barrel of his gun. Another time, he pulled the tail of an iguana, who turned around to look at him square in the face, as if to say "What made you pull my tail?" So much to stir the imagination. Darwin spent only five weeks of a five-year journey there, but roughly one-quarter of his notes and field book are dedicated to this mesmerizing part of the world.

Darwin grasped the significance of his observations after he returned to London, when he set to work on his theory of evolution by natural selection. By the turn of the nineteenth century, people were already struggling to reconcile the Judeo-Christian creation story with the ancient earth accounts put forward by the burgeoning field of geology. Yet it was proving difficult to leave behind the Aristotelian *scala naturae* that meshed so well with biblical notions of human dominion and Enlightenment ideals of progress. Darwin was concerned about the public reception of his radical idea, and it would be years before he accumulated enough evidence to publish. In 1858, more than two decades after the *Beagle*'s famous voyage, Darwin made a joint announcement with the British naturalist Alfred Russel Wallace, who was about to go public with a similar idea. The following year, Darwin published his seminal work, *On the Origin of Species*.

This book rejected the existence of a "natural hierarchy" of any sort. In one fell swoop, Darwin challenged creationist ideas that species were made by God, forever immutable, and qualitatively distinct. It was long assumed that exquisite adaptations (such as the intricate hinges and patterns of bivalve shells) were evidence of a designer at work. Darwin instead showed that species evolve and diverge through processes of natural selection, whereby organisms better adapted to their environment tend to survive and reproduce—helping to resolve that "mystery of mysteries" Darwin once called the origin of new species. All living beings on earth are connected through common ancestry, our minds emerging through evolutionary processes with deep roots in Nature. In his later book *The Expression of Emotion in Man and Animals*, Darwin elaborated on his understanding that if physical traits are evolutionarily continuous, then so too are thinking and feeling. There is no unbridgeable human-animal divide, or impassible barrier between species. The Great Chain of Being is simply a myth. He later wrote to his friend that writing his book *On the Origin of Species* was "like confessing a murder."

Darwin's evidence revolutionized our understanding of our place in the world. It is widely considered one of the great paradigm shifts away from anthropocentric thought. But fast-forward to the twenty-first century and our culture has yet to adopt many of its precepts. We don't like to admit that we are animals. We deny it outright or say and do things incompatible with this uncomfortable piece of information. We've rejected so many other inaccurate models of the world, such as the geocentric universe and the flat earth, but we continue to embrace the *scala naturae*.

Darwin would likely be surprised by human exceptionalism's lingering grip on the public imagination today. But even he foresaw it: "The main conclusion arrived at in this work—namely, that man is descended from some lowly organized form—will, I regret to think, be highly distasteful to many persons." Perhaps we shouldn't be too surprised, then, to find that scientific efforts to assert human exceptionalism redoubled in the decades that followed.

I've often wondered if the implications of Darwinian theory ramped up the human superiority complex. Intriguingly, recent research suggests that human exceptionalism is a worldview we cling to when threatened. The "Ascent of Man" measure was developed by Nour Kteily and colleagues at Northwestern University based on notions of humans rising to the top of a biological hierarchy. In one study, they showed participants the famous image of five silhouettes proceeding from quadrupedal hominid ancestors to bipedal modern humans. Participants then rated their perceptions of the "evolvedness" of various human groups—such as Americans, Arabs, and South Koreans—listed beneath the image. Kteily found that the presence of threats primed participants to assert their in-group (in this case, other Americans) as a hierarchically unique or superior form of life. Kteily's research team studied threat in relation to intergroup conflicts among humans, but perhaps the underlying psychology is shared with respect to other species. The more we are made aware of our vulnerability and threatened by animality, the more we must assert our dominance and superiority. As Melanie Challenger, author of the 2021 book *How to Be Animal*, points out, "This generates a curious paradox, of course, if we perceive being animal as a threat in and of itself." Perhaps evidence for evolutionary continuity made the Differential Imperative all the more imperative.

Moreover, if in drawing a firm line between man and animal, the main purpose of early modern theorists was to justify practices like vivisection, hunting, domestication, and meat eating, these practices (and the need for justification) further proliferated in the industrial era. As industrialization spread and human power to dominate Nature scaled up, human use of other species formed an increasingly essential part of the narrative of human progress. "Progress" was the watchword of this Victorian era, and we know how evolutionary ideas were exploited by Social Darwinists to sanction imperialism, eugenics, and other agendas in the name of humanity's advancement. Darwin himself, I'm afraid to say, even offered at times a racist and sexist view of humankind. In retrospect, the story of evolution—that there is no natural hierarchy of beings—seems so compelling. And yet as history has shown time and again, values often matter more than the evidence. "Despite Darwin," wrote Lynn White Jr., "we are *not*, in our hearts, part of the natural process." Despite Darwin, the quest for human distinction marched on.

Moving the Goalposts

When Jane Goodall first witnessed chimpanzee David Greybeard collecting long stems of grass and using them to fish for termites, the alleged line between humans and other animals suddenly got blurry. The year was 1960, and Goodall—a twenty-six-year-old typist with no formal scientific training—had been sent to Gombe National Park in Tanzania by her mentor, the esteemed paleoanthropologist Louis Leakey. Her observations there would challenge some of our most cherished assumptions about human uniqueness. Among them was the prevailing view that humans were the only species to use tools. Goodall sent an enthusiastic telegram to Leakey, which brought his famous reply: "Now we must redefine 'tool,' redefine 'man,' or accept chimpanzees as humans."

Goodall's observations caused a stir in the scientific community. *Man the Tool-maker* was the title of the critically acclaimed 1949 book written by the anthropologist Kenneth Oakley, who once asserted that "the most satisfactory definition of man from the scientific point of view is probably Man the Tool-maker." *Homo faber*—Latin for "man the maker"—often signifies humans' supposedly unique ability to control their environments with tools. There was thus considerable resistance to the idea that any other species could use tools. Yet subsequent findings showed that Goodall's observations were not one-offs. Chimpanzees and other primates are habitual tool users. Countless other species are also experts in this regard—rats, dolphins, hunting wasps, beavers, and tuskfish, to name just a few. Some of the most astonishing feats, however, come from our avian cousins.

The woodpecker finch, one of "Darwin's finches" on the Galápagos Islands, uses cactus spines or wooden splinters to dig insects out of holes. Egyptian vultures throw stones forcefully at ostrich eggs to break them open. Carrion crows drop walnuts on roads and wait for them to be smashed by cars. The crows are famous for synchronizing their efforts with vehicles at crosswalks. As the traffic light changes and people cross the street, the crows hop out to drop their walnuts, then promptly return to the sidewalk and wait until the light switches again to retrieve the cracked nuts from the road. One of my favorite examples comes from striated herons. These birds gather tiny items (such as bread crumbs, feathers, and mayflies) and drop them into ponds just below their fishing limbs. The items act as bait, attracting small fish into the herons' striking range.

I once overheard two scientists at a conference arguing over tool use in other animals. One proposed that while other species can clearly *use* tools, humans are the only species who can *make* tools. That is to say, unlike other animals, we are unique in our ability to fashion and manipulate tools toward specific ends. I still remember his noticeable discomfort at a session later that day led by researchers studying crows on the Pacific island of New Caledonia. New Caledonian crows, it turns out, commonly bend twigs into hooks and even make serrated rakes from stiff leaves.

In a series of clever experiments, researchers presented bird participants with wires of different shapes, alongside a tiny food-baited bucket sitting at the bottom of a transparent tube "well." In the first experiment, the crows had a choice between hooked and straight wires, and could get the bucket out successfully only if they used the hook. On a later trial of the experiment, a crow named Abel removed the hooked wire, leaving his comrade Betty with just the straight piece. After trying to use this unsuccessfully, Betty wedged and tugged the wire with her beak—crafting a hook!—which she then used to retrieve the bucket. This was not something she did by accident on that one occasion; it was an innovative solution to a novel problem that she continued to implement in subsequent trials. The paper describing these observations shook the field of comparative cognition. Later work showed that tool manufacturing is part of the species' natural behavioral repertoire.

These instances of avian tool use are also noteworthy because birds are relatively far away from us evolutionarily (compared with, say, our close primate relatives, whose intellect we more readily acknowledge), and various ancestral species in between don't appear to use tools. There are many examples of distantly related organisms independently evolving similar traits to solve similar environmental problems (a process known as convergent evolution). Just because human evolution transpires in one way doesn't mean there aren't other ways for the evolutionary story to unfold.

At this point you might be thinking, but humans use tools for all manner of reasons, whereas other species seem to primarily do it to obtain food. However, it's not simply in the service of sustenance; other animals use tools in communication, construction, and recreation. Octopuses gather coconut shells to create shelter and use rocks to form barriers. Elephants pick up branches with their trunk to swat flies, scratch themselves, and even plug up water holes they have dug so that the water doesn't evaporate. In another one of my all-time favorite examples, tree crickets construct acoustic baffles by cutting a hole near the center of a leaf and singing from inside of it to amplify sounds they make for communication during mating rituals. When faced with a choice between two leaves, the crickets follow the same logic principles (e.g., make a hole the size of the wings) that scientists have "discovered" to achieve near optimal results. *Homo faber* we call ourselves, but tool use, it seems, has never been solely our crowning achievement.

Another quality thought to distinguish humans from other species was put forward in 1871 by Edward Tylor, the founder of cultural anthropology. Tylor defined culture as "that complex whole which includes knowledge, belief . . . and any other capabilities and habits acquired by *man* as a member of society" (emphasis added). I wonder what his reaction would have been nearly a century later when Japanese primatologists first documented the cultural practices of other primates. On the island of Koshima, a local schoolteacher witnessed a young female monkey (later nicknamed Imo, meaning "sweet potato") wash a potato in a small stream. The observation was soon shared with the research team of Kinji Imanishi, who was studying the Koshima macaques in an effort to understand the evolutionary basis of human society. One of the defining features of culture is that knowledge and practices must be transmitted socially among the group. Subsequent observations confirmed that Imo's innovation spread through the troop through social learning, both within and across generations (the descendants of these monkeys still wash potatoes today, more than eight generations later). Potato washing provided other evidence for fundamental aspects of culture, including modification (the behavior eventually migrated from the freshwater stream to the sea, presumably adding a salty taste). Reluctant to ascribe culture to other species, scientists originally put the word in scare quotes ("culture"), or used terms like "pre-culture" or "proto-culture," to emphasize that it wasn't "true" culture. It took

Western science decades to catch up, but today there is growing evidence and acceptance that many animals are cultural beings (including insects like bumblebees and fruit flies), though the majority of research remains on other primates. There are even conservation initiatives underway that aim not just to preserve biological diversity but to prioritize this cultural diversity as well. Time and again, the supposed line separating humans from the rest of the living world is unclear.

Still another failed criterion was put forward in the late 1800s by Sir William Osler (a Canadian physician widely described as the father of modern medicine), who argued, "A desire to take medicine is, perhaps, the great feature which distinguishes man from other animals." The subsequent documentation of medicinal plant use in chimpanzees—who swallow rough leaves to rid their intestines of parasites—led to a surge of research in this area across species. For instance, many animals (including deer, tapirs, and parrots) seek out and consume clay, which treats digestive problems by absorbing intestinal bacteria and their toxins. Some lizards are believed to respond to venomous snakebites by eating roots that counter the venom. Fruit flies lay eggs in plants containing high ethanol levels when they sense parasitoid wasps, protecting their offspring. Pregnant elephants in Kenya even eat specific leaves to induce delivery. Anyone who has seen a dog eat grass has likely witnessed self-medication—the grass is thought to aid digestion and help them purge the body of parasites and toxic substances.

Recent studies show that chimpanzees not only self-medicate but also tend to and treat others' ailments, a behavior indicative of empathy. One 2022 report led by Alessandra Mascaro and Lara Southern describes an intriguing behavior performed by a chimpanzee community in Loango National Park in Gabon. The researchers noticed that chimps with open wounds captured insects and then rubbed them along the exposed surface of their injury (it's thought that these insects contain compounds that facilitate wound healing). Occasionally, the chimps applied insects to other group members' wounds. In one striking instance, Southern describes how "an adult male, Littlegrey, had a deep open wound on his shin, and Carol, an adult female, who had been grooming him, suddenly reached out to catch an insect. . . . What struck me most was that she handed it to Littlegrey, he applied it to his wound and subsequently Carol and two other adult chimpanzees also touched the wound and moved the insect on it. The three unrelated chimpanzees seemed to perform these behaviors solely for the benefit of their group member." As we'll soon see, some scientists remain skeptical about the presence of prosocial behavior (any behavior intended to benefit others) in species other than our own. But the link to emotional mechanisms like empathy (sharing, understanding, and responding appropriately to the states of others) tends to raise even more eyebrows.

Shortly after taking Frans de Waal's undergraduate primatology class, I began working as a research assistant in his brown capuchin monkey lab. On my first day, a beautiful female named Winter approached the front of her group's enclosure and lip-smacked at me, a friendly gesture capuchins often do to greet one another. Immediately taken, I instinctively lip-smacked back, and her photo sits in my office to this day, a token of our initial, emotional exchange.

The numerous studies in recent decades providing evidence for emotions in other animals have made it hard to defend outdated Cartesian ideas that they are unconscious automata. Yet we've still been compelled to draw a line somewhere. Many would concede that animals exhibit "primary" emotions like fear, joy, and sadness. But it's often argued that "secondary" emotions, which include empathy, pride, and guilt, exist only in humans, requiring a level of sociocognitive sophistication and self-awareness uniquely attributed to us (for this reason, they are sometimes called social or self-conscious emotions). The Differential Imperative strikes again!

Although skeptics still balk at the term "empathy" when applied to other animals, there is growing evidence that secondary emotions are not what separate "us" from "them." Returning to the image of Nikkie and Luit in the trees, imagine a third chimpanzee—a by-stander uninvolved in the original conflict—approaching the victim and offering some kind of comforting contact. This prosocial act, better known as consolation, is one that Frans (and subsequently I and many other researchers) also observed in chimpanzees. It's now considered the best-documented behavioral marker of empathy in the animal kingdom—reported in species like Eurasian jays, ravens, dolphins, elephants, voles, and beyond.

Recently we ticked another emotion off the list: jealousy. As it turns out, when newcomers to the group arrive and threaten established relationships, chimps try to protect those bonds, similar to how humans act when they are jealous. I was surprised that reviewers of our study didn't seem to mind our use of the word "jealous" to describe the chimpanzees' behavior. Perhaps this is just a sign of the times, a growing acceptance of secondary emotions in other species. But it may also reflect a more pervasive tendency to reserve positive characteristics (like empathy) for ourselves, while readily attributing negative traits (like aggression, malice, and jealousy) to our animal natures. Yet as Stephen Jay Gould reminds us: "Why should our nastiness be the baggage of an apish past and our kindness uniquely human? Why should we not seek continuity with other animals for our 'noble' traits as well?"

And what about rationality, the fateful quality that scholars thought for years was the special province of human beings? *Homo economicus*, we anointed ourselves, pointing to our unique capacity to make rational decisions. Yet the evidence often suggests the opposite. Humans routinely fall prey to bad advice. We take mental shortcuts that result in errors and bias. We deny overwhelming evidence about things like climate change, vaccines, and smoking. Comparative studies have shown that other species appear more rational (at least according to human metrics—a point we'll revisit) in various social contexts. For instance, when learning from others, human children across diverse cultures tend to copy a demonstrator's actions even when these actions are causally irrelevant to solving the task at hand. In contrast, animals like dogs, dingoes, and apes often omit irrelevant actions when there is a more efficient way to solve the problem. Moreover, contrary to predictions of traditional economic models, humans are not rational maximizers in ultimatum games, in which chimpanzees maximize their benefits more efficiently.

These findings have even spurred a countertrend, which holds that human beings are the uniquely *irrational* animal (unsurprisingly, we were quick to reframe this weakness as a strength—a sensitivity to social context, including a penchant for "over-imitation" in the copying studies and fairness in the economic games). Yet several issues complicate the conclusion that humans are singular in their irrationality. Studies show that other animals can be just as irrational as us in certain situations, displaying many of the classic "human" biases. In short, when it comes to reason, it's still unclear what, if anything, sets our species apart.

Descartes suggested that language is the hallmark of a rational mind, and thus of human uniqueness. Many scholars past and present echo this claim. *New York Times* journalist Ferris Jabr put it well: "Language is the last bastion encircling human exceptionalism." Yet even here, we now recognize that other species are breaching the ramparts, challenging whether core linguistic features are exclusively human.

Researchers who long maintained that the distinguishing feature of human language was semantics—where specific signals carry specific meanings—had to reassess when other species were found emitting acoustically distinct alarm calls for different predators. Vervet monkeys gained fame for their ability to encode whether an approaching predator is a leopard (to which they produce a "bark"), an eagle (a low-pitched staccato "rraup"), or a snake (a high-pitched "chutter"). Animal behavior scientist Con Slobodchikoff found that prairie dogs specify the type *and* the size, shape, color, and speed of potential predators in their calls; they even combine the structural elements of these calls in inventive ways to describe entities they have never seen before.

Those who claimed it was syntax—the ordering of words to create limitless meanings—were astonished in 2016 when a group of scientists showed that Japanese great tits exhibit compositional syntax. These birds use one string of calls (ABC) to tell others to scan for danger, and a distinct call (D) to invite them to move closer to the caller. When they hear ABC-D, the birds combine the commands, cautiously surveying the area before approaching the caller.

In the songs of many bird species, we find grammatical structures. Starlings, it turns out, can be trained to distinguish between two types of sound patterns: simple sequences (like a list) and more complex ones with nested (or recursive) elements (in which parts of the sequence are repeated within themselves, like layers). These talkative birds trouble the notion that complex grammatical structures like recursion are unique to human languages.

No doubt, some will point to yet another untested capacity as "the" stamp of language, excluding other species until proved otherwise. And never mind this is by our own linguistic standards; might other species follow their own language rules that remain entirely foreign to us? *Homo loquens* (Latin for "speaking man"), we're fond of calling ourselves, but again the dividing line feels ill-defined.

The Sentence

In his 2006 book *Stumbling on Happiness*, Harvard psychologist Dan Gilbert introduces something called "The Sentence." He writes that various professionals take vows—physicians take the Hippocratic oath to do no harm, priests swear on the Holy Bible to remain celibate. He continues:

"Few people realize that psychologists also take a vow, promising that at some point in their professional lives they will publish a book, a chapter, or at least an article that contains this sentence: 'The human being is the only animal that . . .' We are allowed to finish the sentence any way we like, but it has to start with those eight words."

Gilbert's tone is playful, but all jokes have some truth. The Sentence makes many psychologists, anthropologists, and philosophers I know laugh in uneasy recognition. Gilbert completes it with ". . . thinks about the future," adding to the various ways The Sentence has concluded over the years: tool use, culture, medicine, reason, language, self-awareness, self-control, consciousness, theory of mind, morality, death awareness, innovation, teaching, art, love, and many others. It would be too large a task to inventory all these definitions of what makes us "human," let alone critique them. But isn't the sheer quantity of these hypotheses enough to arouse suspicion? Can each of these be *the* defining characteristic that distinguishes "us" from "them"? And time and again, haven't the hypotheses been disproved? Indeed, Gilbert jokes that most psychologists put off completing The Sentence for as long as they can in their careers, so they might die before being proved wrong.

In a 2018 paper titled "Why Do We Want to Think Humans Are Different?," the biological anthropologist Colin Chapman asks us to consider: "If hypotheses about human uniqueness repeatedly prove to be wrong for one trait after another, does this not imply that the hypothesis itself is wrong? We can keep resurrecting the hypothesis with new traits not yet considered, but to what end?" The answer is clear enough: as we've continually defined and redefined what it means to be human, we've excluded certain groups from moral consideration and justified their mistreatment. And the problems with that have never been limited to other species.

Not *All* Humans

Are women human? This question may sound ludicrous today, but it has been asked in some form many times throughout civilization's history. In antiquity, women, foreigners, and others deviating from the dominant white-male Greek ideal weren't actually conceptualized as people at all. References to human hierarchies abound in Aristotle's *Politics*; women are depicted as inferior to men but higher ranking than enslaved people. It was often argued that these subordinate human groups were primarily constituted by their physical bodies rather than by anything rational. Aristotle (in)famously refers to the female character as "a sort of natural deficiency."

This illustrates a fundamental point: the worldview of human exceptionalism has never placed all humans on a par. Rather, it has historically denied full humanity to certain human groups, subordinating others on the basis of their supposedly animal-like qualities. As with the Differential Imperative, this compulsion to demarcate the "human" has sanctioned exploitation along not only species but also ethnic, racial, gender, sexuality, disability, class, and other lines. It's telling that throughout history, human groups thought to lack the attributes in which other species were supposedly deficient— reason, culture, intelligible language, morality, technology, and the like—were considered "subhuman."

The ethnographic record is teeming with examples of such characterizations. "Beasts in the skin of man," a Jacobean clergyman once said of the Khoekhoe people (an Indigenous population of south-western Africa), their speech "an articulate noise rather than language, like the clucking of hens or gabbling of turkeys." They were "filthy animals," said another traveler, who "hardly deserve the name of rational creatures." According to the English historian Edward Gibbon, speaking of coastal hunter-gatherer societies, "In this primitive and abject state, which ill deserves the name of society, the human brute, without arts or laws, almost without sense or language, is poorly distinguished from the rest of the animal creation." What a profound irony that these human groups were deemed "primitive" because their lifeways were incomprehensible to their colonial oppressors. In 1609 the English writer Robert Gray said that the earth was "possessed and wrongfully usurped by wild beasts . . . brutish savages, which by reason of their godless ignorance and blasphemous idolatry, are worse than those beasts." The Earl of Clarendon similarly proclaimed that "the greatest part of the world is yet inhabited by men as savage as the beasts who inhabit with them." At the turn of the eighteenth century, European colonists hunted down the Indigenous peoples of New England with dogs; "they act like wolves and are to be dealt withal as wolves," stated one clergyman as justification.

In Great Chain of Being conceptions, not all humans were created equally in God's image. A rigid hierarchical social order decreed by God justified divides between peasant and noble classes and granted absolute power to monarchs (a.k.a. the divine right of kings). "The numerous rabble that seem to have the signatures of man in their faces," explained Sir Thomas Pope Blount in 1693 of the poor, "are but brutes in their understanding. . . . 'Tis by the favour of a metaphor we call them men, for at the best they are but Descartes's automata, moving frames and figures of man, and have nothing but their outsides to justify their titles to rationality." Time and again, non-white, non-Western, and other human groups deviating from the dominant prototype are characterized as lacking or deficient in refinement, culture, rationality, self-restraint, and morals, dependent on instinct and not reason, and more tolerant of pain. Gerald O'Brien, a professor of social work, has analyzed the frequent "animalization" of people with cognitive disabilities: "Their reportedly high procreation rates, their inability to live cultured lives, their presumed insensitivity to pain, their propensity for immoral and criminal behavior, and their instinctual rather than rational nature" have all contributed to their being excluded from the "human" community. Nazi propaganda portrayed Jews as pests, advocates of slavery depicted Africans as apes, and Europeans likened the Romani people to "a race of vermin." Dehumanization scholars like David Livingstone Smith have argued that this language not only accompanied colonization, slavery, and extermination, but facilitated these atrocities.

Animal insults remain a feature of human discourse today. When animal terms ("pig," "rat") are used metaphorically to refer to humans, their meaning is usually derogatory. Donald Trump makes a habit out of it, calling former White House staffer Omarosa Manigault Newman a "dog" on Twitter and deploring unauthorized immigrants: "These aren't people. These are animals." More often, the tendency is less explicit, operating beneath conscious awareness. Experimental research has shown that people implicitly associate members of their out-group, relative to their in-group, with "animal" over "human." And no doubt even these unconscious biases have real-world consequences. One set of studies by psychologist Philip Goff and colleagues found that American adults retain an implicit association between apes and Black humans, endorsing violence against Black suspects and affecting decisions in criminal justice contexts. Using archives of actual criminal cases, researchers found that news articles written about Black (versus white) people convicted of capital crimes were more likely to contain ape-relevant language, and that this portrayal was associated with a higher probability of receiving death-sentencing judgments.

Is it any coincidence that the word "cattle" shares the same etymological root as "chattel"? Or that the word "mulatto" shares its etymology with "mule"? Or that "race" and "husband" emerged from terms for animal breeding?

Every time I have taught The Arrogant Ape, without fail, some version of the following question comes up early in the semester: How can we attend to issues of human exceptionalism toward other species when we can't even manage to achieve justice for members of our own? That is, how can we even dream of addressing cruelty toward other living beings given all the horrible things we do or let happen to fellow humans every day? While these questions are perfectly understandable, they overlook something central. Namely, that various forms of discrimination are interconnected and rooted in a broader, usually tacit acceptance: a concept of the human that is set apart from the rest of the living world on a false hierarchy of being.

This doesn't mean equating or even directly comparing different forms of discrimination. Decolonial theorist Aph Ko provides a helpful illustration in her work on the relation between racism and speciesism. She likens it to a grammar system. If you imagine that white supremacy is a sentence, racial violence would be represented by one word in that sentence and animal violence by another word. The two forms of violence are separate words, but you need them *both* to make sense of white supremacy. Ko maintains that because racism and speciesism are so tightly linked, current movements for social justice and animal rights cannot be fully realized when pursued in isolation.

Consider also the role that species hierarchies play in mainstream human rights discourses, which often condemn practices on the grounds that they violate human dignity by treating people "like animals." As one example, in the 1995 Eighth Amendment case *Madrid v. Gomez*, the treatment of prisoners at California's supermax Pelican Bay State Prison is repeatedly likened to the treatment of other animals. *But what does it even mean to be treated like an animal?* Why would being caged or forcibly restrained seem any more fitting for other animals than for human beings? Canadian philosophers Lisa Guenther and Will Kymlicka have argued that these animalized discourses fail to adequately describe (and thus challenge) the harm of prison programs because they remain complicit with a hierarchical opposition between human and animal. We might condemn treating prisoners like dogs, Guenther explains, but "we cannot fully understand the brutality of these programs until we refuse to accept that dogs deserve to be treated this way, any more than humans do."

More and more research has found that the psychological factors that underlie human exceptionalism serve to reinforce and promote prejudice against humans—a finding known as the "interspecies model of prejudice." On the flip side of this coin, reducing the status divide between humans and other animals can help combat discrimination and strengthen notions of equality among human groups. Studies show that beliefs in evolution (and namely, perceived human-animal similarity) predict lower levels of bigoted attitudes and discriminatory behaviors against members of LGBTQ, Black, and immigrant communities, even after controlling for religious and political views. The relationship is not merely correlational but also causal. One set of studies by Kimberly Costello and colleagues had participants read newspaper editorials that emphasized either human-animal similarities or differences. Those who learned about human-animal similarities reported lower prejudice against immigrants and other marginalized human outgroups than the participants who learned about the human-animal divide. The interspecies model of prejudice has been replicated even among children: the less children are taught to place the human above the animal, the less they dehumanize racial minorities. In short, and often contrary to popular opinion, attention to human exceptionalism need not detract from issues of human justice. Instead, it can help us better understand the common roots of these issues and confront them with greater clarity.

⸻

To recap, Western intellectuals have long been preoccupied with human distinction. From at least 500 B.C. to the present, the so-called Differential Imperative has sparked endless proposals of traits that make us special and, ultimately, more morally worthy. What all such proposals have in common is that they assume a duality between "human" and "animal" and invariably regard the latter as inferior.

Did this compulsion to assert human superiority arise in response to harmful practices toward others that demanded rationalizations? Or was it the other way around, where human exceptionalism gave rise to those very practices in the first place? As with all chicken-and-egg problems, it's impossible to say. While it's certainly true that we invoke the human distinction to justify exploitative systems toward others, it's hard to imagine that these systems would have ever been tolerated in the first place if these others had been credited with fully human attributes.

Some of the Differential Imperative's earliest architects were the progenitors of modern philosophy and science. Though many of their ideas have been discredited today, it is important to recognize that they established the terms of a debate about the minds of other species that continues to shape our thinking well into the present. Today, evolutionary continuity among all lifeforms is widely accepted in relation to biology (anatomy, genetics, development, and so forth), but this view remains controversial when it comes to the mind, and especially consciousness. Are humans undoubtedly more cognitively capable than other species? Many people would still answer definitively yes. But, as we shall see next, credible comparisons depend on comparable conditions.

CHAPTER FOUR

Pride That Blinds

M ORE THAN EIGHT MILLION PRIMATES LIVE IN NEW YORK City, and surprisingly, some of them aren't human. Hundreds if not thousands of monkeys live in Manhattan alone. You can find them at the Central Park Zoo. Or shacked up in apartments as illegal pets, prompting neighbors to dial 311. But if you're part of the animal scientific community, you know that many of them are kept lawfully in research laboratories, carefully hidden away from the public eye.

Every morning I would ride the subway to 168th Street in West Harlem, emerging from the underground turnstiles to a dark brick building towering overhead. I'd take the elevator to the ninth-floor annex of the New York State Psychiatric Institute, where a dozen singly housed rhesus macaques awaited my arrival. The elevator bell would chime and, like clockwork, they'd erupt into loud barks and screams, shaking their cages and spinning in rapid, repetitive circles (scientists call this "stereotypical behavior," a sign of stress). They calmed down once they realized it was me and not a member of the veterinary staff coming to cart them off for a medical exam or procedure. I was studying their numerical cognition—whether they had mental categories for counting and ordering like humans do. I'd transfer them one by one to stainless-steel test chambers across the hall, where they'd interact with computer touch screens for an hour a day for food rewards.

From the moment I set foot in the building, I knew these macaques didn't lead good lives. These tiny cages, this sterile environment, the social isolation, their monotonous daily routine—all while the city below pulsed with energy, vibrancy, and life. I was instructed not to get too close to the cages, lest a monkey lash out and grab my hair, clothing, or a finger. So I dreaded going into the lab. Every bone in my body rejected it. Other lab assistants, most of them like-minded "animal lovers," were also there to gain experience in primate research. We occasionally discussed our discomfort with this kind of captive work. One of them quit (ultimately leaving science behind altogether) because the research was so far removed from the respect for other species she'd envisioned this field being all about. I tried to quell my own emotional anxiety. It felt like another rite of passage, one I myself barely made my way across. But it was one of the few opportunities I encountered for students interested in animal behavior science, and I felt one step closer to being able to study their minds and interests.

Stacking the Deck

This lab was my point of entry into Columbia's PhD program in psychology. Formally the science of mind and behavior, psychology is often defined more narrowly as the scientific study of the *human* mind and behavior. But even with more inclusive definitions incorporating other species, some people consider the basic questions psychologists ask useful mainly insofar as they relate to our own species' psychology and evolution. Ask any primatologist, and they will tell you that grant funding often hinges on how much the proposed research will promote the scientific understanding of humankind!

It was during my graduate studies that I became familiar with the field of comparative psychology. Comparative psychologists study the behavioral and mental processes of different species side by side—most often comparing the cognition of humans with that of our closest living relatives, the great apes.

Some comparative psychologists study physical cognition: how individuals use and acquire information about the physical world, comprising topics like spatial cognition, timing, and numerical competence—as was done at the macaque lab where I worked. Other researchers focus on social cognition: how individuals use and acquire social knowledge, including topics like theory of mind (the capacity to understand others' knowledge and beliefs) and prosocial behavior (any behavior intended to benefit others). Comparative studies often claim human superiority in physical and social cognitive function. But too often, the deck is stacked against other species, giving humans an unfair advantage from the start. Let's consider primate prosociality, a topic close to my heart and one I've studied for years.

Classic experiments on prosocial behavior compare how humans and captive chimpanzees respond to another individual in need. For instance, studies on helping might assess whether participants try to retrieve out-of-reach objects for experimenters. Other studies on altruism test whether they share resources (like food) with other individuals, including strangers. Such experiments have generally concluded that humans are exceptional in their motivation to help and cooperate with others. By comparison, the cooperation of other primates is thought to be mainly limited to kin, and virtually never extended to unfamiliar individuals. According to some, this merely reflects animals acting in their own self-interest, since the behavior benefits their own genetic relatives. But there's a catch. Several, actually.

For one, the characteristics of the study populations are not representative of the two species. The captive chimpanzees have typically spent their lives with a single, small social group in highly stable, restricted human-made environments. Such situations could not be more different from those of wild chimpanzees. In the wild, chimps live in flexible societies characterized by "fission-fusion dynamics," meaning that the size and composition of the social group change as time passes and animals move through their environment. Wild chimpanzee groups can contain thirty to one hundred individuals of all ages at any given time, and they inhabit large home ranges with territories of at least seven square miles. In contrast to captivity, where chimps are frequently separated from their biological mothers at birth, wild chimp mothers stay with offspring for extended periods and have an enormous influence on their development.

The comparison group comprises free-living humans, often WEIRD children living in their natural homes and with their families, who come to the research institutions solely to participate in the study. You may recall that despite being the most common subjects of psychological research, WEIRD people are unrepresentative of humanity on the whole. We know that cognitive processes—theory of mind, spatial ability, causal reasoning, and others—vary markedly across human cultures. Yet research frequently takes WEIRD people as the norm, and study designs often reflect WEIRD assumptions. For instance, comparative psychology experiments attempting to measure altruism in humans and great apes often assume that sharing should be the preferred strategy over not sharing when there is no cost to oneself. Yet this assumption doesn't even hold in all *human* populations: in some cultures, sharing implies a contractual obligation. But blanket statements ("humans are more altruistic than chimpanzees") are still commonplace, with little concern for whether the study represents the target populations—humans and chimpanzees—on the whole.

How can we routinely turn a blind eye to these inconsistencies? In my view, conclusions about species differences, at least based on study designs like this, can reveal only a hidden bias of human exceptionalism at work. Blinded by our commitment to this worldview, we overlook how the game is rigged against other species—undermining their abilities, and our ability to understand them.

~~~~~~~~~~

The summer after I began research in the monkey cognition lab, I decided to try my hand at fieldwork. Luckily, a baboon project affiliated with the University of Cape Town was seeking research assistants. I applied, got the job, and several months later boarded a flight to South Africa. The field site was located on the Cape of Good Hope peninsula in Table Mountain National Park. The area boasts it's "where two oceans meet" (the Atlantic and the Indian Oceans). In truth, the oceans collide a bit farther southeast. But for accommodation totaling twenty rand a night (about $1.10), which guaranteed a cot in the lighthouse technicians' quarters, waking up to plunging coastlines all around roused no complaints from me. It was a magnificent deal.

We were involved in monitoring a troop of chacma baboons only partly habituated to human presence. This meant following them around at a distance from dawn to dusk and collecting data on where they went and what they did. I learned here that a day in the life of a wild monkey could not be more different from that of a captive one. Food has to be located and extracted. Neighboring troops and predators loom. Sleeping sites must be scouted out and secured before dark. My first week felt like a form of baboon boot camp. These animals were constantly on the move, up and down hilly terrain, maneuvering acrobatically through dense brush, while I stumbled to keep pace. The weather on the peninsula could change in an instant. Rain-bearing winds made for sudden torrential downpours. The baboons would mysteriously take cover under a rocky outcrop in the nick of time, with me—wringing wet—following suit moments too late. I often wondered whether the baboons were prescient, or whether city life had made me woefully unattuned to my natural surroundings, or (as I ultimately concluded) whether it was a bit of both.

Given the unique coastal topography of this region, these baboons have learned to supplement their diets with protein-rich foods like mussels, limpets, and shark eggs—exemplifying their remarkable behavioral and dietary flexibility. Marine foraging behavior fluctuates depending on environmental factors like the tide, swell height, and wind (winds frequently exceed gale force, knocking me off my feet on various occasions). All of this was subject to marked seasonal variation throughout the year. They lived in an ever-changing environment. Each day was different.

Complex foraging strategies alone require cognitive adaptations related to orientation, mapping, prediction, and causal reasoning. But consider the cognitive demands of their social lives, too—keeping track of friends and foes, rearing infants, learning from others, managing conflicts, coordinating activities, communicating, and beyond. It made the barren NYC annex feel more like a prison than a lab (. . . and for what crime, exactly?). I couldn't bring myself to go into the lab anymore after I returned from that trip. For ethical reasons that were already clear. But it was becoming all the more apparent for scientific reasons as well.

Ample evidence exists that different environmental experiences affect the cognitive development of all kinds of species. Decades of research in human psychology tell us that impoverished environments have deleterious effects on mental function. Studies on prison inmates, for instance, reveal significant cognitive and emotional deficits as a result of incarceration. It would thus be unthinkable to represent human cognition based on this population. So why doesn't a similar logic apply to other species?

Yet to this day, animals raised in unnatural, barren, or restrictive environments are commonly considered exemplars of a species' cognitive capacities. One 2018 study by Martin Glabischnig of the University of Amsterdam found that most of what gets cited in major review and theory papers on primate cognition comes from captive (as opposed to wild) populations. In another analysis, primatologist Christophe Boesch reviewed a decade's worth of publications in top scientific journals that directly compared human and chimpanzee performance on cognitive tasks. Strikingly, *all* such studies contrasted the performance of apes in captivity with that of free-living humans.

Ecological Validity

A laboratory environment can rarely (if ever) adequately simulate the natural circumstances of wild animals in an ecologically meaningful way. Comparative studies thus often lack "ecological validity," a measure of how findings generalize to real-world contexts for the species in question. Thus, should it come as any surprise that studies show humans outperforming laboratory chimpanzees on various sociocognitive tasks? However, in the wild, entering *their* worlds, the picture looks quite different.

Recall prosociality studies showing that captive chimpanzees (in contrast to WEIRD human participants) don't appear to share food or work as a team toward joint goals, with cooperation restricted mainly to kin. Yet in the wild, chimpanzees share food with unrelated group members and work cooperatively during hunting and risky intergroup encounters quite frequently. My friend and colleague Liran Samuni has observed this directly in Taï National Park,

Côte d'Ivoire. She found that chimps participate in highly coordinated group-hunting activities, which involve cohesive search efforts for prey and subsequent meat-sharing. Sharing is especially prevalent among hunters, but not limited to them. Most adult bystanders (70 percent) also receive meat, highlighting that sharing is widespread and not just beneficial to those who participate in the hunt or their relatives. More generally, wild chimps share not only meat but also fruit, nuts, honey, and other resources like tools, and the strongest predictor of this behavior is not kinship but the presence of enduring and mutually preferred social bonds. Some researchers have even argued that the majority of mutual aid in the forest comes from unrelated chimpanzees. Observations in natural environments thus suggest ape-human differences in cooperation are not as stark as usually proposed.

None of this challenges the fact that captive studies have revealed some major facets of chimpanzees' cognition. We just can't assume an individual's abilities in these environments transfer to other contexts. Unlike their captive brethren, wild animals have autonomy in their lives—when to eat, travel, and rest and with whom to socialize. They make these choices and learn from their consequences. They live in complex and often unpredictable environments, continuously adjusting their behavior to changing temperatures, humidity, daylight hours, food availability, social situations, and other factors. Their cognition is an adaptation for dealing with these challenges.

These days I conduct research only in the wild and in sanctuaries. But I've worked in all kinds of settings—including primate labs and zoos. My colleagues and I have found not only that chimpanzees console each other following conflicts (a behavior indicative of empathy) but also that individuals vary markedly in this tendency. In other words, some chimps are consistently more empathic than others. Research in the wild also suggests that the behavior may not be universal. One chimpanzee community living in the Mahale Mountains of Tanzania appears to console, while another in the Budongo Forest of Uganda does not (or does so incredibly rarely). This points to the exciting idea that beyond material cultures (population differences in behaviors like tool use), chimpanzees also have social cultures (population differences in social norms and tendencies), as is the case with humans. We are now testing this possibility by comparing consolation rates across multiple groups of chimpanzees at Chimfunshi Wildlife Orphanage, a forested sanctuary in Zambia. The presence of group variation in social tendencies like empathy would further challenge blanket claims about chimpanzees in general. Too often, we look at chimpanzees as a "species" rather than as the members of diverse communities and individuals they actually are. Why should this nuanced outlook be reserved for humans alone?

My latest work investigates whether chimpanzees also reassure other group members in distressing situations beyond conflicts, including after the death of a loved one. One day, my student Zoë Goldsborough arrived at the zoo to find an adult female chimpanzee named Moni holding a stillborn infant, whom she had given birth to the night before. Moni had only recently been introduced to this group and didn't have many social ties yet. Since she was a newcomer, we guessed that her pregnancy had been particularly stressful, as was the stillbirth. But that day, we saw the group interacting with her in ways we had never seen—following her closely, grooming her, and embracing her gently. They were attempting to reassure her, providing the first clear evidence that other animals console bereaved group members. We have yet to see whether wild chimpanzees engage in similar behavior toward the bereaved, but I'd be willing to bet they do.

Beyond resource sharing, we know that unrelated chimpanzees in the wild regularly provide coalition support to allies and jointly patrol their territory borders to fight against rival groups, a risky feat with potentially high costs including injury or death when partners don't cooperate. There are numerous instances—both in the wild and in captivity—of chimpanzees adopting unrelated orphans. New examples of helping behavior arise all the time (recall from the last chapter the recent study showing that wild chimpanzees engage in wound care of not only themselves but others), challenging humanity's special claim on kindness. These cannot be explained through "selfish" motives like kinship. Yet despite all this, today it is still debated whether other species, and chimpanzees in particular, are capable of behavior not motivated by self-interest.

Spot the Difference

There are additional ways we rig the game against other species in these comparative contexts. As the comparative psychologist David Leavens has highlighted, "All direct ape-human comparisons that have reported human superiority in cognitive function have universally failed to match the groups on testing environment, test preparation, sampling protocols, and test procedures." To illustrate, I show my students a video on a set of studies measuring altruism in chimpanzees and human children. I ask them to compare the methodologies for measuring helping behavior in each species—sort of like those "spot the difference" games where players must find the discrepancies between two seemingly similar images. They immediately note how humans are usually tested with conspecifics (i.e., other humans), while chimps are tested with members of another species (i.e., humans). Human infants and children typically remain in close proximity to their caregivers during the task (often sitting on the lap of, or in direct visual contact with, a parent), whereas chimps are separated from the group (not to mention often from their mothers from a young age). Humans are almost always tested in the same area as the experimenter, while chimps are typically separated by some sort of physical barrier like steel bars or plexiglass walls. This means human participants can readily receive subtle forms of encouragement from experimenters or caregivers, which are usually absent to the chimpanzees. There is one notable similarity, which actually ends up being a difference: humans are tested with conspecific tasks and materials, whereas chimps are typically tested with the same (i.e., human!) tasks and materials.

The list of potential confounds is endless (one of my students kindly pointed out how my "spot the difference" metaphor was misleading, because there are more differences than similarities).

Researchers defend these methodological discrepancies by saying that studying humans and chimpanzees in identical ways is impractical (for instance, it can be dangerous to have chimpanzees and human experimenters in the same room). Most of the time, however, such practical limitations don't prevent comparative psychologists from making strong claims about other animals' shortcomings and humans' cognitive uniqueness. Yet all along, what people have identified is an effect of different physical and social environments rather than species differences.

Scientific Speciesism

To make this point even clearer, consider addiction research. For decades, experimental psychologists studied human addiction using rodents in Skinner boxes—small boxes or cages that usually contain only a bar or lever that an individual animal can press to gain a reward, such as food, or avoid a painful stimulus, such as a shock. Researchers would give rats the opportunity to self-administer small doses of a drug by pressing the lever. They found that rats would press the lever frequently, consuming large amounts of heroin, morphine, cocaine, and other drugs. The research seemed to prove that rats—and by extension, humans—with access to these drugs become addicted. The conclusion was one on which the mass media quickly capitalized. The message that drugs are irresistibly addicting fit the dominant narrative at the time propagated by antidrug campaigns, politicians, and companies (the "war on drugs" was popularized by news cycles in 1971 after a press conference given by President Richard Nixon in which he declared drug abuse "public enemy number one").

In the late 1970s, Canadian psychologist Bruce K. Alexander challenged conventional wisdom about drug addiction with a set of experiments colloquially known as the "Rat Park" studies. He constructed a large plywood box and filled it with things that rats enjoy—platforms for climbing, running wheels and balls for recreation, wood chips for burrowing, and plenty of food and mating space. Unlike traditional rat cages, Rat Park contained lots of rats living and playing together.

Alexander and his team then ran experiments comparing the drug intake of rats living in Rat Park and those kept in conventional laboratory cages. In virtually every experiment, the Rat Park residents consumed less of the drug solution. And not just a little less, *a lot* less (in one case, nearly twenty times less). These results suggested that the addictive behavior of rats living in Skinner-like boxes was— at least in part—an artifact of their impoverished physical and social surroundings, not merely the accessibility of the drugs themselves. Today, scientists appreciate that drugs may have addicting properties, but that environments and psychology are also crucial factors to consider in understanding the full causes of addiction.

Alexander has drawn intriguing parallels between his Rat Park findings and narratives around the addictive behavior of colonized human groups in Canada. Initially, English colonizers framed the addictive behavior of Indigenous peoples as evidence of their biological inferiority—namely, that they were drawn to alcohol, drugs, and gambling because of some genetic vulnerability to addiction. But as with the Rat Park findings, addiction appears limited in the natural environment and increases dramatically when individuals are placed in contexts producing social and cultural isolation. "In both cases, the colonizers or the experimenters who provide the drug explain the drug consumption in the isolated environment by saying that the drug is irresistible to the people or the rats," Alexander clarifies. *"But in both cases, the drug only becomes irresistible when the opportunity for normal social existence is destroyed."*

The evolutionary biologist Stephen Jay Gould exposed similar interpretive, racist biases in human intelligence research. For years, studies on IQ purported to demonstrate that certain human groups—namely white Western Europeans—were innately superior in intellectual ability. But such studies almost always failed to account for political and social factors like systems of housing, education, criminal justice, health care, and virulent racial oppression. In doing so, they promulgated a biased, racist worldview under the pretense of science, something widely known today as "scientific racism."

Gould also noted the profound lack of consistency in sampling criteria and testing protocols in intelligence research (standardized testing is still criticized today for taking knowledge more common to certain human groups as the benchmark of intelligence). His point was not that intelligence researchers were intentionally unscientific, but rather that their judgment was clouded by the expectation of hierarchy, leading them to collect and interpret data in biased ways. These researchers fell prey to something psychologists call confirmation bias. Because they were committed, a priori, to the idea of essential differences between different "kinds" of people, they turned a blind eye to the inconsistencies in their own research.

A parallel systemic blindness exists in contemporary comparative cognition research. That is, when comparing humans with other species, differences in cognitive performance are often assumed to reflect differences in innate ability rather than environmental and other methodological variables. These assumptions are rarely explicitly articulated by researchers, but they are nonetheless pervasive in comparative cognition and evolution research on the whole. People have long co-opted the authority of science to confirm their ideas of racist and sexist human hierarchies. The same can be said of species hierarchies. I call it *scientific speciesism*. Scientific speciesism is another confirmation bias, leading researchers to seek or interpret information in a way that's consistent with existing, anthropocentric beliefs about the world. Fortunately, when we become aware of these biases and start to see how they impact results, we do better science, and learn more about other species and ourselves in the process.

————

In the late 1940s, the psychologist Donald Hebb wanted to compare the intelligence of rats raised in a sterile research laboratory to that of rats raised in a complex, stimulating environment. So he brought a bunch of rats home to live among his family. (I like to picture Hebb walking in his front door, hanging his coat and hat like any other day, and then—rather nonchalantly—unleashing a crate of rats who scurry in all directions to make themselves at home.) Later, when he ran them through mazes, he discovered that his pet rats were much smarter than his lab rats.

Since then, many studies have demonstrated that placing animals in enriched environments positively impacts their cognition. Compared with those reared in captive environments, wild-born chimps show superior cognitive functioning on standard measures such as learning speed, discrimination, and understanding functional relationships. These patterns persist even after twelve years of environmental enrichment for the captive chimps, suggesting that the deficits incurred are long-term and not readily corrected. Similar results have been reported in many other animals under human confinement and control—paramount among them, farmed animals.

In their natural habitat, cows grow up in complex social groups. In contrast, dairy industry calves are typically separated from their mothers at birth and raised in individual pens. Recent studies have found that compared with isolated calves, those raised in a group not only have better welfare but also perform better on learning and memory tasks. Likewise, hens have higher well-being and cognitive performance when they live in aviaries than in cages (the two standard rearing systems for laying hens in the European Union). This same general phenomenon—enriched environments yield happier, smarter individuals—has been reported in pigs, goats, sheep, salmon, and many other farmed species. Sometimes findings have conservation applications. Research by Victoria Braithwaite and Anne Salvanes showed that when cod in standard hatcheries were exposed to environmental enrichment, they developed enhanced behavioral traits associated with improved survival in the wild. This is especially important because captive-bred hatchery animals are sometimes released into their natural habitats as part of a conservation strategy to restore threatened populations. But these reintroduction efforts are controversial and often fail, in part because the barren environments in which these farmed fish typically develop ill-equip them for life in the complex wild.

As I realized during my first field season in South Africa, a world away from the sterile NYC primate lab, experiencing a dynamic environment is what allows animals to thrive and develop flexible behavior and cognition.

Losing Their Minds

Years ago I attended an animal ethics workshop in Switzerland, a country known for having some of the strictest animal testing laws in the world (a recent campaign even aimed to make it the first country to ban animal experimentation altogether, which lost in a 2022 referendum). At the workshop, a neuroscientist on the defensive argued that he would happily come back in his next life as one of the mice kept in his research lab. After all, they have regular access to food and water, protection from predators, and refuge from extreme weather. Their basic needs were met, he insisted, and they were all the healthier and happier for it. I've heard similar accounts for zoo animals—who "get" to spend their days lazing around, eating, and socializing. But as many of us know firsthand from the COVID-19 lockdowns, monotony and boredom (not to mention restricted autonomy) take a serious toll on our physical and mental health. Why would that apply only to the human species? I'll never forget a meme that went viral several months into the 2020 quarantine—a photograph of an orca in a tiny circular tank that read, "Next time you're upset about being in quarantine, imagine . . . 50 years of swimming in circles!" It struck a chord with me, and a million others.

Like us, other species learn best through curiosity and free exploration, not through restricted, homogenous environments in which they have little agency. Studies demonstrate that acute challenges are good for animals' well-being and cognition. The problem is when they are subjected to artificial environments and unavoidable manipulations, leading to repeated stress and anxiety. As with humans, chronically stressed individuals lose their cognitive potential and emotional capacities (anhedonia).

But strangely enough, sometimes stress is actually the point. Take the Morris water maze, a task widely used to investigate spatial learning and memory. The task forces rats to frantically swim in a water tank with high walls until they find a platform to alight on (in subsequent trials, researchers test their memory of the platform's location). Then there's the Columbia Obstruction Method, in which animals deprived of basic needs—such as food, water, or social interaction—must traverse an electrified grid so that researchers can test whether their motivation to reach sustenance or a mate outweighs their fear of getting shocked. Today we wouldn't dream of testing humans under such conditions. We see it as obviously unethical. But it's also unscientific, because we know that severe stress compromises cognitive performance. Think about it: Imagine you were asked to take *any* test under severe threat. What is the possibility that your test results would accurately reflect your abilities?

Yet despite this, at the monkey lab where I worked, it was common practice to restrict macaques' diet to boost their motivation on the cognitive tests we gave them. Indeed, many labs keep their animals at 85 percent of typical body weight to help ensure food motivation. But what's really unnerving is when that practice results in claims about human specialness. Take one 2017 study titled "Universal and Uniquely Human Factors in Spontaneous Number Perception," which concluded that humans were superior to rhesus macaques in various facets of numerical cognition. Tucked away in the fine print of the methods is that the monkeys in the study were kept on a water-restricted diet. Imagine depriving a child of food or water before subjecting them to a series of math tests. Would this be a just way to measure their numerical ability?

Recently, neuroscience demonstrated that living in an impoverished, stressful environment physically damages the brain. These changes have been documented in practically every species studied to date. When researchers look at the posthumous brains of animals like captive orcas, dolphins, rodents, elephants, and primates, they find reduced synaptic connections, shrunken capillaries and neurons, less-complex dendrites, and other changes that disrupt the brain's complex circuitry. For many species, living in physically and socially restricted environments seems to thin the cerebral cortex in particular—the part of the brain thought to be involved in "higher" cognitive functions like memory, planning, and decision-making. It's one of many reasons why zoos and aquariums have come under heightened scrutiny lately (as noted in a *Wall Street Journal* article aptly titled "When Animals Lose Their Minds"). What blinds us from seeing the pitfalls of studying animals with negative psychological experiences and aberrant brain development? What keeps us from understanding that doing research in this way jeopardizes the relevance of the science?

The same questions apply to other fields. In U.S. laboratories alone, more than 115 million animals—including mice, rats, frogs, dogs, cats, rabbits, hamsters, guinea pigs, monkeys, fish, and birds—are killed annually for scientific research. Most serve as models for human health, a practice that many people (like that neuroscientist at the Swiss workshop) still vehemently defend. They say it would be immoral to do the same invasive work on humans. But the logic is paradoxical: other species are close enough to serve as models for human biological and psychological disease but different enough to fall beyond ethical concern—a contradiction possible only, I might add, in a culture dominated by human supremacy.

And despite decades of animal research, scientists have made very few inroads to treatments for human psychiatric and other illnesses. Success rates in human clinical trials are staggeringly low (currently one in nine drugs entering human trials succeeds). There is mounting concern over the "replication crisis," in which many findings in fields like psychology and medicine don't hold up when others try to repeat the study. People are increasingly suspicious that the failure to translate from animal research to human outcomes may come from how we approach animal research itself. After all, as Stanford medical scientist Joseph Garner points out, "every drug that fails in humans 'worked' in an animal model." A core assumption in biomedical studies is that animals used in experiments embody healthy biological and psychological systems. But most live in enclosures that are an infinitesimal proportion of their average home range—a mouse cage is typically 280,000 times smaller than a mouse's natural range, or seven million times smaller in the case of a rhesus macaque. Research on laboratory rodent health and survival reveals that the standard housing for these animals should be considered a severe stress condition. Today we have the technologies to study free-roaming animals in the wild or in captivity under more ecologically valid conditions, not to mention viable and more ethical alternatives to animal testing through computer simulations and *ex vivo* human models (which often prove more effective than animal models in the first place). When we recognize that other animals are minded beings shaped by their experiences and deserving of moral consideration, don't *we too* become all the healthier and wiser for it?

Illusions of Control

A popular notion is that laboratory experiments allow one to control for the influence of possible confounding factors. By reducing subjects' biological and environmental variability—a process known as standardization—we can more readily discern how our chosen variable influences the outcome of interest. But more and more research shows that standardization does not necessarily make us more confident in our results. The literature is replete with papers reporting poor reproducibility of findings, hence the replication crisis noted earlier. A result that is highly reproducible under highly controlled contexts often generalizes poorly to other environmental contexts—a problem known as the "standardization fallacy." Let's consider the role of context further with the following question: Can honey bees see in color?

The famous German ophthalmologist Carl von Hess studied color vision in honey bees by placing honey bee workers in experimental chambers, where environmental factors could be experimentally controlled. He presented them with two light spots simultaneously, each varying in color and light intensity. He repeatedly found that the bees were attracted to the brightest spot, no matter the color. In 1912,

Sir von Hess (who was previously knighted for his work on the science of vision) published the first comprehensive book on color vision in animals. In it, he concluded that all invertebrates, including honey bees (and even fish), were color-blind. It sparked a heated debate and a historic academic rivalry.

That same year, the young Austrian ethologist Karl von Frisch boldly challenged his eminent colleague's claims. If existing pollinators like honey bees can't perceive color, von Frisch reasoned, then why do flowers come in all sorts of different colors? Von Frisch trained and tested his bees in their natural habitat, placing them in situations where he expected they'd be motivated to learn and to display their ability to distinguish colors. At his country home, he set up a table in an area sheltered from rain and direct sunlight, and gave honey bees a sugar solution in a small glass bowl atop a colored card, set among an array of other cards in various shades of gray. As von Frisch himself describes:

"By the scent of a little honey it is possible to attract bees to an experimental table. Here we can feed them on a piece of blue cardboard, for example. They suck up the food and, after carrying it back to the hive, give it to the other bees. The bees return again and again to the rich source of food which they have discovered. We let them do so for some time, and then we take away the blue card scented with honey and put out two new, clean pieces of cardboard at the site of the former feeding place—on the left a blue card, and on the right a red one. If the bees remember that they found food on the blue, and if they are able to distinguish between the red and blue, they should now alight on the blue card. This is exactly what happens."

Von Frisch successfully demonstrated that honey bees see color in the behavioral context of feeding and homing, a finding that von Hess mercilessly refuted. But as it turns out, von Hess was not entirely wrong: bees *are* color-blind in their escape runs toward the light, the context in which he studied them. He just incorrectly generalized that bees are color-blind in *all* behavioral situations. Being color-blind in one context does not indicate an overall inability to perceive color (humans, for instance, are color-blind in dim light or darkness, hence the expression "all cats are gray at night"). This illustrates two fundamental points. First, studies in natural environments are far more likely to reveal what other species are actually capable of, because they tap into the ecological pressures they've evolved to navigate. Second, when it comes to understanding these capacities, we'd be mistaken to generalize findings obtained in highly controlled laboratories to other environmental contexts.

More and more researchers today appreciate that an individual's abilities are shaped by interactions between their evolutionary history and their developmental experience. This recognition recently prompted some to advocate for the equivalent of WEIRD in other animals—including BIZARRE (barren, institutional, zoo, and other rare rearing environments) and STRANGE ("S" for "social background"; "T" for "trappability and self-selection"; "R" for "rearing history"; "A" for "acclimation and habituation"; "N" for "natural changes in responsiveness, genetic makeup"; and "E" for "experience") to denote the suite of factors that can lead to sampling biases in animal research. Yet the paramount role of experiences is still overlooked, which is why some continue to controversially assert that the cognitive capacities of animals like laboratory chimpanzees are not negatively impacted by their unnatural environment.

So where might this assumption—that other animals' cognitive capacities are minimally affected by their environments—stem from? As you may have guessed, it's one deeply embedded in Greek, Christian, and Enlightenment conceptions of human supremacy. For instance, recall Descartes's contention that humans are thinking beings uniquely endowed with souls, while other animals are automata lacking in all experience. Unlike humans, other species merely respond to stimuli by following a predetermined program, much like a machine. Or as Christophe Boesch puts it, "Humans are born resembling a white sheet upon which everything can be written, whereas animals are genetically rigidly fixed." A monkey is a monkey. A rat is a rat. A honey bee, a honey bee. No matter where they're from.

Yet for an integrated biological entity—a living being—life experiences and environments are paramount. Assuming that would be the case only for humans illustrates a bias toward human exceptionalism—that humans are thinking and feeling subjects shaped by experiences, while other species are objects that develop the same under a whole array of environmental conditions. While these Cartesian notions are much criticized today, they are strikingly persistent in many subtle ways. To understand why, we need to look back to other scientific traditions that gave them legs.

The Behaviorist Carpet

At the end of a long day, under a saturating sunset sky, the baboons began to settle in to a sleeping site. As dusk fell, so did their activity levels, making my life a little easier. No longer struggling to keep up or record data, I found myself sitting still, silently watching the spectacle of their lives unfold around me. These were the moments I enjoyed most.

Three juveniles play on some rocks in the distance, one chasing the other until they tumble into a furry ball. An exasperated mother tries to wean her infant, causing him to throw a theatrical tantrum ("How *could* you!"). Nearby, two sisters groom each other with a sense of urgency as daylight recedes. Suddenly, I hear a loud "wahoo" call behind me, portending an aggressive conflict between two males.

Dozens of baboon eyes dart toward the threat, except for the two groomers'—they triple the intensity of their grooming, as if busying themselves might keep them out of harm's way. If you pay attention, a drama unfolds every other minute in baboon society. And just like a drama, there's no shortage of intrigue, desire, ulterior motive, and emotion. It is impossible to adequately capture the scene before you without invoking the mental lives of these animals in some way.

The rise of behaviorism in the early twentieth century attempted to banish all mentalistic concepts from the field of psychology. The premise was that mental phenomena are unobservable, making their existence unverifiable. According to one of behaviorism's leading proponents, B. F. Skinner (recall the Skinner boxes in which rats repeatedly pressed a lever to receive drugs), the goal was "to predict and control the behavior of the individual organism." All actions were thought to be governed by simple stimulus-response rules, often conditioned via reward and punishment, without appeal to inner thoughts or feelings. Behaviorists thus portrayed other animals as mechanical, not so unlike the image of the animal-machine inherited from Descartes in the seventeenth century. Brushed under the behaviorist carpet, animal emotions and other mental states were dismissed out of hand, and inferences about them deemed dangerously unscientific.

By sweeping animal minds under the rug, behaviorists departed dramatically from early naturalists like Darwin, who readily understood animal life as subjectively meaningful. So convinced was Darwin about the rich inner worlds of other animals that their capacity for attention, memory, imagination, and other states of consciousness (even dreams!) struck him as plain common sense. He used animating language that accorded complex motives and emotions to other species, exemplified by the following account of male birds showing off their plumage:

"Ornaments of all kinds . . . are sedulously displayed by the males, and apparently serve to excite, or attract, or charm the females. But the males will sometimes display their ornaments when not in the presence of the females, as occasionally occurs with grouse . . . and as may be noticed with the peacock; the latter bird, however, evidently wishes for a spectator of some kind, and will shew off his finery, as I have often seen, before poultry or even pigs. All naturalists who have closely attended to the habits of birds are unanimously of opinion that the males delight to display their beauty."

After my first year of graduate school, having resolved to move away from laboratory research, I started reading more studies in ethology. Ethology was partially forged in opposition to the behaviorist school of comparative psychology and emphasizes the evolutionary basis of animal behavior under natural conditions. The pioneers of ethology—Konrad Lorenz (as I recently learned, a member of the Nazi Party), Niko Tinbergen, and Karl von Frisch (the bee researcher mentioned earlier)—did Nobel Prize–winning work illuminating the behavioral and communicative patterns of other animals. But even they were skeptical of the possibility of engaging in a serious scientific study about animal minds. In contrast to Darwin's appeal to inner states and intentions to describe animal behavior, classical ethological writing often depicts behaviors in mechanistic terms. Genes dictate behavior via instinct. Behavioral patterns are "activated," "released," and "triggered" by the environment. Organisms don't act on their surroundings; *they are acted upon*. Drawing on Darwin's account of peacock behavior above, Eileen Crist compares the general approach of ethologists with that of earlier naturalists:

"When the same behavior of the peacock is considered from a classical ethological standpoint its import changes radically. . . . He is compelled to execute a fixed action pattern by inner-physiological forces that are beyond his control, and, of course, outside the pale of his understanding. . . . The peacock's odd display is no longer a misfired performance, but something he cannot help doing. The behavior looks empty, meaningless, and mechanomorphic—in a word, *mindless.*"

But why should *mindless* be considered the mark of a rigorous scientific explanation?

A Cardinal Sin

Crist contrasts these "mechanomorphic" descriptions with the "anthropomorphic" language endorsed by scientists like Darwin. Anthropomorphism originally referred to the heresy of projecting a human form onto God. Today, anthropomorphism typically refers to the projection of human characteristics onto other species, and ethologists like Tinbergen warned vehemently against it. Early in my own career, I learned that anthropomorphism was a cardinal scientific sin—akin to doing bad science. Some caution is of course warranted. Just because other animals behave in similar ways doesn't mean we should assume it's for the same reasons. A classic example is the facial expression commonly believed to be a "smile" in chimpanzees (lips pulled back exposing the top and bottom teeth). Those seemingly happy chimps you sometimes see in TV commercials and films are actually displaying a fear grimace, something this species does when afraid or distressed. Ham "the astrochimp"—the first great ape to travel to space, who appears to smile in photographs but is obviously terrified—has become an unfortunate poster child for why anthropomorphic interpretations can be so misleading.

But somewhere along the line, accusations of anthropomorphism took on a decidedly different tone. That is, to assign "minds" to other species in any meaningful sense was to indulge in a kind of anthropomorphism. The outdated religious and philosophical background of this position is hard to deny. In reality, fears of this kind of anthropomorphism often hide a pre-Darwinian mindset, one that rejects mental continuity between species. When scientists balk at anthropomorphic explanations, however, they instead point to Morgan's Canon, which suggests we should not attribute complex psychological processes to animal behavior when simpler explanations will do (think of it as a kind of cognitive parsimony). But more often than not, assuming similar mental processes between humans and other species *is* the simplest explanation for what we observe (also known as evolutionary parsimony). I can't tell you how many times I've witnessed the mental gymnastics required to *not* attribute a thought or emotion to another animal; it seems so much more straightforward to assume continuity between species!

Once, an editor asked my coauthors and me to remove the word "empathy" from the title of a paper on chimpanzee consolation. They preferred to restrict the observation to the behavior itself, given that we can't be sure that chimpanzees who console are motivated by empathy. I can follow this line of reasoning, but only up to a point. We have evidence across various species that the behavior serves to reduce the recipients' distress, occurs most often among close social partners, and is regulated by a bonding hormone called oxytocin— all of which are consistent with an empathy hypothesis. But above all, the exact same behavior expressed in children *is* considered empathic, and usually without question. I therefore prefer the more parsimonious explanation: that other species think and feel in ways with which we can identify. Thanks to human exceptionalism, however, we are usually reluctant to admit it. Frans de Waal coined a term for it, "anthropodenial": "a blindness to the human-like characteristics of animals or the animal-like characteristics of ourselves."

Finally, is it really anthropomorphic to call chimpanzees empathic, and a form of anthropodenial otherwise? As the philosopher Brian Keeley has argued, calling such claims anthropomorphic implies the trait is intrinsically human and only derivatively nonhuman (instead, he suggests that traits shared with mammals might be mammalomorphic, for example, or primatomorphic when shared among all primates). That is, even when anthropomorphic interpretations appear warranted, labeling them as such—as human-like— assumes that *humans* have a premium on the capacity in question. In Keeley's words, "to draw such a conclusion is to commit a fundamental misunderstanding of the biological world and the place of humans within it." As we've seen, that misunderstanding stems from Great Chain of Being conceptions of the universe, where *Homo sapiens* continuously enjoy a special claim on characteristics like sentience and intelligence.

Fortunately, many scientists today agree that nonhuman animals possess a variety of subjective states—that they experience the world in intelligent and sentient ways. There are now countless scientific and popular books on this very topic. It seems a new, groundbreaking discovery about animal minds is made each day. And not just for those with whom we readily identify, like other primates. Evidence is now amassing for the emotional and cognitive worlds of species to whom we are more distantly related—birds, crustaceans, and insects. Yet despite all this knowledge about other species, we evidently still underestimate the depth of their psychology, which compromises our research methods. According to philosophical ethologist Dominique Lestel, this bias creates a self-fulfilling prophecy: by removing animals from their meaningful physical and social environments, we effectively transform them into reactive machines, and then study them as such.

Self-Fulfilling Prophecies

Lately, I find myself worrying about self-fulfilling prophecies. Studying animals in restricted, human-made environments and then drawing conclusions about their abilities is one thing. But I also worry about how this unfolds in the increasingly human-dominated spaces we sometimes call "the wild."

As human populations grow, more and more species will live in environments characterized by severe anthropogenic disturbance. The Food and Agriculture Organization of the United Nations estimates that the last three decades have seen 420 million hectares (more than a billion acres) of forest lost and converted to human land use. We already know that this endangers populations and biodiversity. But might forcing other animals to live in a world of our own making also imperil behavioral and cognitive diversity? A 2019 *Science* study led by Hjalmar Kühl provided evidence for this very idea. One of the most important outcomes of decades of research on wild chimpanzees has been the discovery of cultural differences in behaviors—hunting strategies, tool use, feeding techniques, social interactions (as briefly noted earlier), and more. Kühl and colleagues analyzed a large dataset of 144 chimpanzee communities who inhabited areas with varying human impacts. Chimpanzee communities in areas with high human impact had substantially (88 percent) reduced behavioral diversity than those in low-impact areas. As the authors argue, in destroying and fragmenting animal habitats, we're not only depleting resources but also disrupting social learning processes crucial for behavioral and cultural transmission. In short, our planetary actions jeopardize both the health of other species and (very much relatedly) their complex abilities—and ultimately our ability to appreciate them.

We tend to focus on losses to biological diversity, such as extinction rates. But do we appreciate that glyphosate, one of the world's most commonly used herbicides, is interfering with the navigational abilities of honey bees and other pollinators? That chemical pollutants are disrupting the shoaling and social recognition capacities of creatures like killifish? That anthropogenic noise (such as that from road and boat traffic) is imperiling vigilance, communication, and migration in species ranging from whales to birds? One particularly telling study on adult zebra finches found that traffic noise hampers spatial memory, motor learning, inhibitory control, and social skills. We are destroying not just biodiversity but species' minds and gifts.

This is not to say that many species don't find ingenious ways to adapt to living in human-dominated habitats. Anthropogenic activity may promote the emergence of cognitive and behavioral innovations. But when we force them to live in a world of our own making, we risk losing sight of who they really are. As a consequence, we lose Nature as a partner and as our best teacher.

The negative consequences of our actions pertain not just to other species but also, of course, to us. Chemical and noise pollutants are highly harmful to our biological health and our ability to concentrate, learn, and remember. Fossil-fuel combustion is driving indoor carbon dioxide toward levels known to be harmful to human cognition. Lead levels have likewise been shown to be associated with deficits in various realms of cognitive and academic performance.

We light up the night, disorienting those animals attracted to light and those who rely on stars to chart their course. But it also makes it impossible for us humans to stargaze, to situate ourselves in the vast universe, perhaps the most humbling activity of them all.

And as long as we're dealing with self-fulfilling prophecies, I also wonder how many of the wild animals we deem dangerous to humans might simply be responding to the trauma they have undergone in their own lives. Many bears in North America have seen their mother shot or snared. Elephants in Africa display symptoms of PTSD as a result of poaching, culls, and habitat loss. We say they are dangerous, but do you blame them? (And who are we to judge?)

When we divide the world into feeling, experiencing subjects on the one hand and objects for our exclusive use on the other, we perpetuate this self-fulfilling prophecy. This myopic approach makes it such that we lose sight of our place in the grander scheme of things. We become alienated not just from the living world but from ourselves.

Looking back, I think working in the primate cognition lab was possible only because, through powerful and often hidden processes of enculturation, I had internalized a bias of human exceptionalism. On some level, it was simpler to suppress my discomfort and tell myself that this was just another means to an end: with a PhD in hand, one day I'd be able to help them. I was assured that this kind of captive control makes for better science. And I wanted to do good science (ironically, to showcase and understand the mental lives of these animals, something these very conditions precluded). And so scientists try to cultivate a state of ethical indifference to their study systems—to distance themselves from the beings with whom they work. But more often than not, the animals find a way to break through this artificial barrier.

Sometimes a nervous male named Macduff would press his body right up against the front of his cage as I walked by. He wanted to be touched, desperate for some form of social contact, so I broke the lab rules and groomed him through the wires. The second I began, he held his body still as a statue, as if the slightest movement might cause me to walk away. Moving my fingers along the fur of this other living, breathing being, I swear I could sense both of our heartbeats slow (that mutual benefit, a sure sign of our shared primate ancestry). I'm pretty sure he would've let me groom him all day. Macduff taught me a lot in those moments—about him and other macaques, about myself and other humans, about connection, empathy, trauma, guilt, and maybe even the potential for forgiveness. We always wore surgical masks around the macaques because diseases pass easily between us close primate relatives. Every time I put on a mask during the COVID-19 pandemic, the sensation brought me straight back to Macduff, and those haunting mornings in Manhattan with the monkeys.

CHAPTER FIVE

The Mismeasure of All Things

ONE OF THE FIRST ONE HUNDRED EPISODES OF *THIS American Life* celebrates the remarkable abilities of pronghorn antelopes. Considered the best endurance athletes in the world, pronghorns can run the equivalent of a marathon in forty minutes, metabolizing oxygen at a rate three and a half times higher than even the finest human runners do. But my favorite feat is tucked away in the program, almost like a footnote to their astounding athleticism: these antelopes are said to have 10x vision, which suggests that, on a clear night, they can see the rings of Saturn.

Umwelt

Jakob von Uexküll, a German biologist, introduced the concept of the "umwelt" in the early 1900s. The umwelt means the world as experienced by a particular organism—the aspects of the environment that one actually perceives and senses. Far from imperceptive machines, all species orient themselves and develop in complex relation with their surroundings.

Antelopes' eyes are located on the sides of their head, affording 300-degree vision. This is a good thing when you're a prey species and need to spot predators in time. You and I have forward-facing binocular vision. Our eyes evolved more to help us navigate and locate food than to look out for our own safety. But it's birds of prey who have the most impressive long-distance binocular vision in the animal kingdom, and I know this firsthand. Every morning in Namibia, the moment I awoke, a southern white-faced owl perched in a tree overhead looked down and locked eyes with me through the mesh canopy of my tent. Accustomed to surveying the ground for small rodents and scorpions, owls can perceive minute movements that would be imperceptible to us, especially in low-light conditions. The mere twitching of my limbs or parting of my eyelids instantly got her attention. She knew I was awake before I did. A staring contest would ensue, becoming a morning routine of sorts between us. What a gift it is, to have a daily ritual with an owl.

Most humans have three types of color-sensing cells in their eyes—attuned to red, green, or blue light. Many birds have a fourth type, sensitive to ultraviolet (UV) light. What looks like a plain yellow flower to us has a UV bull's-eye in the center for them, guiding them to the pollen. A 2020 study led by Princeton ecologist Mary Caswell Stoddard revealed that hummingbirds may see an even greater variety of nonspectral colors than previously understood. They can combine widely separated parts of the light spectrum (such as red and UV light), meaning they can probably see entire dimensions of colors we can't even imagine.

Harlequin mantis shrimps are thought to have the most sophisticated visual system in the living world. With sixteen different types of photoreceptors in their eyes (twelve for color analysis), they can detect a special spiraling type of light called circularly polarized light. Some species of scallop have up to two hundred eyes, inside of which are mirrors made of tiny crystals that focus light onto the retinas, much like a telescope. Scientists don't know exactly why they evolved, but these compound eyes are already paving the way for novel bioin-spired optical devices. We often chalk technologies like this up to human ingenuity, but the design principles were discovered and honed by Nature long ago.

Other species rely not on sight but touch to navigate their worlds. Star-nosed moles, who reside in near-complete darkness in North American bogs and wetlands, are functionally blind. But they have ultrasensitive snouts—some 100,000 nerve fibers concentrated in a one-centimeter area. That's five times more touch sensors than in the human hand, compressed into a region smaller than a fingertip!

Catfish, for their part, *taste* their surroundings. Their entire bod-ies are covered in taste receptors (up to 175,000 taste buds; most hu-mans have fewer than 10,000), which helps them hunt in murky waters with low visibility, sensing the flavors of potential prey from all directions.

There's an infamous elephant named Charles who lives in Na-mibia's Hoanib River Valley. Charles has earned a reputation for dig-ging up the underground water pipes that supply the nearby tourist camps, though a watering hole is easily accessible nearby (elephants apparently enjoy the shower they get from a broken water pipe). Time and again, the staff have tried to relocate and hide the pipes, but time and again, Charles outwits them.

Elephants can sniff out water sources from miles away thanks to their spectacular sense of smell. I remember once being lucky enough (or so I thought) to spot Charles while on foot. He appeared as a mere speck through my binoculars, so I assumed I was at a safe dis-tance. But the animal tracker I was with told me that Charles had clearly picked up our scent downwind and was shaking his head and ears in a way that communicated agitation. This was one of my first real-world lessons about the importance of understanding other ani-mals' umwelten. Some of the most innovative conservation solutions also rely on this understanding, as exemplified by research on human-elephant conflict. Across their range in Africa and Asia, elephants habitually raid the crops of the human communities near which they live. Visual barriers like fences typically fail to keep the elephants out. More successful mitigation strategies have taken elephants' extraordinarily sensitive smell into account, deploying olfactory deterrents such as chili ropes (which tend to be more cost-effective to boot).

Unsurprisingly, elephants also have remarkable hearing, allowing them to communicate with infrasounds well below the range of human hearing. On the other side of the spectrum, greater wax moths can hear pitches up to 300 kilohertz, fifteen times higher than the frequencies perceptible by the human ear. Owls, too, possess extremely sensitive hearing; they can reportedly hear the heartbeat of a mouse twenty-five feet away. Is it possible, then, that the white-faced owl above my tent also listened to mine? We live in a magic, otherworldly world.

Dolphins, bats, and other species who echolocate experience sound as three-dimensional. They emit high-pitched, ultrasonic calls beyond the range of human hearing, the echoes of which allow them to construct an acoustic sense of their surroundings. With this superpower, echolocating dolphins can perceive what's inside the objects they approach. A dolphin swimming toward you could likely sense your internal organs and bones, just as she discriminates between the fishes that she hunts on the basis of the size, shape, and geometry of their swim bladders.

There is a wealth of other senses outside the standard five. Sea turtles, migratory birds, and butterflies can detect earth's magnetic field, allowing them to navigate over vast distances. Snakes like vipers, pythons, and boas can perceive infrared (thermal) radiation. Platypuses, sharks, and teleost fish have electroreception, meaning they can sense electric currents. Platypus bills may even combine pressure and electric fields, giving these marvelous creatures what scientists think may be a kind of "electrotouch."

Von Uexküll was especially interested in time perception: in how "the length of a moment varies in different animals." Interestingly, the limits of human visual temporal sensitivity make motion pictures possible. A film is just a long strip of still images ("frames") projected at a speed at which we see smooth, continuous movement rather than a frame-by-frame slideshow. That speed is typically around twenty-four images per second, because at that rate, our eyes and brain fuse the images. But other species have different sensitivity. Consider instead animals very unlike us. Consider snails.

Relative to humans, snails move very . . . very . . . slowly, and their perception of time is correspondingly sped-up—meaning they miss movements that we can perceive. Events must be more than a quarter of a second apart for the snail to perceive them as distinct. In the snail's world, waving a finger in front of them four times per second would appear as a single, stationary finger. Sloths, as their name betrays, are also notoriously slow (three-toed sloths have the slowest metabolism of any mammal on earth), which may rev up their perception of time relative to ours. The sloth researcher William Beebe mused in 1926: "Sloths have no right to be living on this earth, but they would be fitting inhabitants of Mars, whose year is over six hundred days long." Beebe was captivated by the ways in which sloths apparently "stretch" time.

Research has shown that animals with fast metabolic rates—like house flies—perceive more information per unit of time, meaning they experience action more slowly than those (typically, larger-bodied) animals with slower metabolisms, including humans. Relatively speaking, they live in a slow-motion, matrix-like world, which may explain their ability to deftly evade swats from fed-up animals like us.

Time perception matters for other sensory modalities, including sound. Many birds produce vocalizations much more quickly than we can perceive—an acoustic richness that remains out of reach to us. Some birds can hear sounds separated by less than two-millionths of a second (whereas we can probably hear two sounds as distinct if they're around two-hundredths of a second apart). The bird gets in one second what it might take us nearly three hours to hear.

These findings show that differences in how organisms perceive the world are not arbitrary but rather finely tuned by interactions with their surroundings, biology, and evolutionary history. Every species has access to a different slice of reality—what is real to one may be invisible to another. There's a vast world around us that other species can perceive but we humans can't. Our umwelt is just one among millions.

The Unfairest of Them All

It may seem like the umwelt concept would be central to research with other species. Yet surprisingly, we often adopt a human point of view and design studies to accommodate human perception. One of the clearest cases of this human-centric bias comes from research on self-awareness. Self-awareness is broadly defined as conscious knowledge of oneself as distinct from others (also sometimes known as self-recognition or self-consciousness).

For decades, we thought that humans were special in our capacity for self-awareness. We concluded this based on results of the mirror self-recognition task, originally developed by Gordon Gallup and colleagues in the 1970s. In the classic experiment, a researcher surreptitiously places a mark on a subject's body. The experimenter then observes whether the subject notices the mark upon viewing their reflection in a mirror. Most human children pass the mark test by three years of age, but almost every other species fails. Bingo! We humans win.

But not so fast. The mirror mark test favors visual species like humans and other primates. What about dogs, who rely primarily on their sense of smell to navigate the world? Other studies have used a novel design based on the dogs' umwelt, extrapolating from their sensory worlds and natural behaviors. Experimenters presented dogs with various canisters, each of which contained an odor. In the key test, they presented dogs with either an "olfactory mirror" of the subject—the dog's own urine—or one in which the odor was slightly modified. Researchers found that dogs could easily distinguish between the olfactory "image" of themselves and other odors. In other words, they exhibited self-awareness.

Only by moving away from the human-centric perspective and taking a dog's-eye (or -nose) view of the world did we get the measurement right. Yet despite widespread acceptance of the significance of smell to dogs, olfaction is typically neglected in canine cognition research. Strikingly, a recent review of experiments in this field revealed that 74 percent of studies published in the last century relied on visual-test paradigms instead.

As the celebrated American poet Mary Oliver once wrote:

> *A dog can never tell you what she knows from the*
> *smells of the world, but you know, watching her, that you know*
> *almost nothing.*

When we overrate skills that are human-like, we miss the incredible diversity of cognitive adaptations that exist all around us. Imagine the possibilities—what we might learn about other species and ourselves—if we took a less human-centric approach. If we grounded our understanding of other species in their umwelten and ways of dealing with the world rather than in how they measure up to our own.

Today, self-awareness is considered a rare capacity in the living world—confirmed in a small albeit growing number of species—likely due to the complex cognition thought to underlie it. As we've learned, the mirror self-recognition test ignores that most species are attuned to themselves less through visual images and more through smell or other sensory modalities. But might it be a restrictive way of understanding self-awareness for other reasons? A squirrel recently taught me an important lesson in this regard.

In Cambridge, Massachusetts, I live in a carriage house surrounded by what a diplomatic neighbor once called an "untamed" garden. But aside from making the pollinators and me happy, the garden offers an amusement park of twigs and trunks for the many gray squirrels who inhabit the area. Last July, a female squirrel was getting plumper by the day, and it soon dawned on me that she was pregnant. It was pretty amusing (if not a little sad!) watching her renegotiate the branches. But before long, she adjusted her behavior. She started picking launch spots with narrower leaping distances and avoiding flimsier limbs in order to stick her landings. Eventually, she found a new route to the neighbor's yard, ignoring the usual gap in the fence through which she could now barely squeeze. These behavioral changes—which are easy to miss or regard as trivial—now made me wonder: Doesn't this require some form of self-awareness? An embodied, corporeal sense of self?

Until recently, body awareness—the ability to take into account one's own body's relationship with the physical environment—was overlooked in discussions about animals' self-awareness. This is ironic, as researchers in human child development have long argued that body awareness is a core component of self-awareness. Research using similar paradigms as those used to study children recently revealed that other animals, too, exhibit a kind of bodily awareness. Diverse species—elephants, dogs, rat snakes, and various insects—consider their own size and shape to solve tasks or safely navigate through complex environments. For instance, in one recent study, researchers showed that bumblebees take into account their own body size both in preparation for and while traversing small gaps (like the expecting squirrel I observed).

Such findings invite us to think about self-awareness more broadly. After all, doesn't a bat adjust his flight path and distinguish the echoes of his own clicks from those of nearby bats? Don't dolphins have "signature whistles," unique vocalizations that function much the same way as names, broadcasting the identity of the whistle owner to conspecifics? Beyond a narrow human-centric lens, we come to see the world more clearly—a world rich in diverse forms of awareness. The question isn't whether other animals are intelligent, but rather, the question Frans de Waal asks in his aptly titled book: *Are We Smart Enough to Know How Smart Animals Are?*

In sum, an anthropocentric approach often creates the appearance of limitations in cognitive ability. It's another form of scientific speciesism—the bias in the way we study other species that confirms beliefs in human exceptionalism and limits valid comparisons of human and animal cognition. The troubling thing is that we have been practicing this bias for so long that, by and large, we are scarcely aware that we are doing it anymore. As we gaze narcissistically into our own reflection ("Mirror, mirror on the wall . . .") and ask ourselves questions with praiseworthy superlatives ("who is the fairest/wisest/kindest of them all?"), we lose sight of the variety and complexity of other beings all around us. The mirror test is only one kind of task for one kind of ability, but its lessons can be applied much more broadly. It turns out that even when we study species with umwelten similar to our own—including our close primate cousins—we still tend to take a far too human perspective.

A Matter of Perspective

After I left Columbia's primate cognition lab, I had to consider carefully whether I was in the right place. In many respects, I was. As the writer Joan Didion famously put it: "Quite simply, I was in love with New York." But I was also in love with fieldwork—where a warm shower was the real luxury and "nightlife" consisted mostly of stargazing or storytelling around a smoky fire. These two extremes—major urban centers like NYC and remote field conditions—are still where I feel most at home. It's the middle—suburbia and all it entails—that feels foreign and disorienting to me.

I was contemplating leaving graduate school because there was no obvious path toward working with other animals, at least in the way I wanted to. So I set up a meeting to discuss these concerns with the director of graduate studies, Dr. Tory Higgins, a well-known researcher in motivation and social psychology. His office, drenched in sunlight, had all the markings of professorial life—overflowing bookshelves, a hatstand in the corner, and a sunken old sofa, perfect for animated discussions with his graduate students.

As I would do on so many days that followed, that morning I entered Tory's office feeling uncertain and left with the conviction that I might be onto an important scientific discovery. I showed him the photograph of Nikkie and Luit, the two reconciling chimpanzees. Together we marveled at our closest primate relatives' complex social lives, and the conflict-resolution strategies they used to maintain them. But, we wondered, were the motivations underlying these strategies also shared across species? Animal behavior often initially sparks curiosity about what it can reveal about humans. Yet, flipping the lens, could we borrow concepts from the relatively well-developed field of human psychology to better understand other animals' minds and interests in these post-conflict contexts (without forcing human characteristics onto them)? Rather serendipitously, these topics coincided with upcoming themes in his lab—close relationships and forgiveness in humans. I called the "Higgins Lab" home for the next five years, and there I explored the underlying factors that motivated reconciliation in humans and chimpanzees. Fortunately for me, these questions were considered legitimate by Tory and the department. But they were also questions that the field more generally had been warming up to.

Recall that under the influence of behaviorism, minds were thought to lack empirical status. Fields like ethology and comparative psychology were among the final outposts of this way of thinking. But in the last half of the twentieth century, the cognitive revolution reintroduced the mind as a credible focus of study. Cognitive ethology—a field concerned with the influence of mental processes on animal behavior—arose at the turn of the twenty-first century and led to an outpouring of research in this area. Since then, more and more researchers have sought to describe the complex mental lives of other animals and to be genuinely scientific about it.

Today most scientists agree that many other animals possess minds and that those minds are amenable to scientific study. Yet as I would come to learn throughout graduate school, hypotheses still tend to consider human cognition the maximum and standard capacity, against which other species can be compared and measured. Protagoras's ancient motto "Man is the measure of all things" still haunts our thinking millennia later. It's a dangerous anthropocentrism that makes every form of life appear deficient by comparison (and paradoxically, doesn't put certain assumptions about human cognitive uniqueness to a fair test!).

Consider the capacity for perspective-taking itself. This ability to understand others' knowledge, desires, and beliefs is sometimes treated as the cognitive Rubicon separating humans from other animals. To test this, studies have often required other animals to use human communicative cues—such as pointing and gazing—to solve a task like locating hidden food. Such studies have typically concluded that animals like chimpanzees fail to consider the visual perspective of others, something at which most humans across cultures excel from a young age.

For instance, in one study, chimpanzees could beg for food from two human experimenters. One of the experimenters could clearly see the begging chimpanzee. The other experimenter, for various reasons (including having a bucket over his head), could not. The study found that chimpanzees did not reliably discriminate between these two experimenters, suggesting they do not adopt another's visual perspective. However, begging from a human (not to mention one unexplainably sporting a bucket hat!) is far from a natural situation for a chimpanzee.

This led the evolutionary anthropologist Brian Hare and colleagues to develop a more species-relevant task. They placed a dominant and a subordinate (lower-ranking) chimpanzee into competition over food—with some pieces of food visible to both and some visible only to the subordinate chimp. In this more naturalistic competitive situation, the subordinate chimpanzee most often pursued the piece of food hidden from the dominant's view! That is, the chimpanzees knew what others could see and even understood what others had seen. It took a new and more ecologically appropriate context for us to realize that chimpanzees can readily discern what other chimpanzees do and do not know.

It's ironic (to the point of being a little funny) that humans would tout superior perspective-taking skills based on experiments that don't take other species' perspectives of the world into account. But as we've seen, the quest to understand what makes our species so special hasn't much shame. A 2022 *New York Times* article—"Is Geometry a Language That Only Humans Know?"—suggested that our ability to recognize geometric shapes is what makes us unique. The article covered a recent study comparing twenty captive Guinea baboons from a research facility in the South of France with free-living humans (French children and adults plus adults from rural Namibia). Researchers showed these baboon and human participants six quadrilaterals on touch screen devices and had them select the one that was unlike the others. The humans consistently outperformed the baboons, leading to the authors' conclusion that "intuitions of geometry are present in humans but absent in baboons." But how representative are these samples? And how relevant is this computer task for the baboons? Might humans live in environments where right angles matter much more than they do in other animals' environments? Maybe baboons have their own intuitions and perspectives when it comes to geometry. Maybe we could even learn something from them.

Mathematical abilities are more likely to reflect a species' way of experiencing and being in the world. Desert ants navigate—via an internal device like a pedometer—by counting the number of steps they take. African lions and spotted hyenas estimate the number of conspecifics before interacting based on numerical odds. And as far as geometry is concerned, I enjoyed reading the comment section of the *New York Times* piece noted above, especially this guy's reaction: "Ever seen a cheetah calculating the correct angle for cutting off a fleeing impala, or a dog calibrating its velocity and speed to plot the most efficient line for intercepting the parabolic path of a tennis ball in flight? Seems like geometry to me. But I'm just a simple country boy."

Wonderfully Different

Each winter, a member of the crow family known as the Clark's nutcracker retrieves food he has previously cached in up to six thousand separate locations spanning over a hundred square miles. In this respect, his memory by far exceeds the average human's memory.

Meanwhile, homing pigeons, using precise internal compasses and landmarks, can return to their lofts from distances exceeding a thousand miles. Their navigational skills have been admired for millennia; by 3000 B.C., Egyptians were using them as a kind of early airmail. There are untold examples of other species showing capacities superior to ours for such complex tasks as memory, spatial learning, causal reasoning, rationality, cognitive flexibility, and others. The more we learn, the more we appreciate that cognitive complexity takes manifold forms in our world, and that other lifeforms exhibit many different kinds of intelligence, some of which humans do not possess.

The cognition of living beings has been crafted through millions of years of evolution. Just as physical traits (like giraffes' necks, which give them advantages in reaching high-up leaves that other animals can't access) are understood in this evolutionary context, so too should cognitive traits be understood in their evolutionary, species-specific context. Life on earth comes in all sorts of shapes and sizes— winged and footed, clawed and quilled, fleshy and finned—each adapted to different environmental niches. Likewise, it's a world inhabited by multiple intelligences, each adapted to solve for different environmental problems. On strictly evolutionary grounds, it is impossible to establish a hierarchy of these intelligences. As the philosopher Martha Nussbaum puts it, "Life-forms don't line up to be graded on a single scale: they are just wonderfully different."

In one of my favorite cartoons, seven animals—a bird, a monkey, a penguin, an elephant, a goldfish, a seal, and a dog—are lined up before a human instructor. "For a fair selection," declares the instructor, "everybody has to take the same exam: please climb that tree." Of course, the same standards of assessment aren't appropriate across species, because each species fulfills a different cognitive niche. There are multiple intelligences! It's therefore not very meaningful to ask whether a human is smarter than a chimpanzee, or a chimpanzee is smarter than a dog, or a dog is smarter than a zebra fish. We're just different—wonderfully different, to echo Nussbaum.

And yet, the definition of "intelligence" we have used for so long—that is, "what humans do"—is incredibly persistent. It's sometimes disguised as a celebration of other species' cognitive capabilities. "Smarter Than You Think" begins a 2010 headline from *The Sunday Times*: "The Animal Kingdom Is Home to Much Greater Intelligence Than Has Been Previously Acknowledged, with Scientists Seeing Evidence of Human-like Traits Everywhere."

Interest in proving animals are clever in human-like ways goes back at least to Darwin, and it's not altogether bad. It is as valid a scientific pursuit as any other (though it may be worth noting that we don't often go around looking for gorilla or dragonfly characteristics in ourselves or in other species). Plus, the more we attribute human-like qualities to other animals, the more we may begin to recognize our own animality. But other species can't be accurately understood, let alone celebrated, if we start from the question of how much like us they are.

And so human cognition often gets put in a league of its own, mysteriously exempt from the typical evolutionary forces to which all other species are subject. "Human intelligence is a biological mystery," opens a 2016 piece in *The Economist*. "Evolution is usually a stingy process, giving animals just what they need to thrive in their niche and no more. But humans stand out. Not only are they much cleverer than their closest living relatives, the chimpanzees, they are also much cleverer than seems strictly necessary." The title of a 2019 *Live Science* article is even more to the point: "Humans May Be the Only Intelligent Life in the Universe, If Evolution Has Anything to Say."

To put it slightly differently: humans are alone in possessing intelligence, because it's a certain kind of intelligence we're after.

Yet if I asked you "Who is more intelligent, Stephen Hawking or the Dalai Lama?" you might rightly wonder: "What kind of intelligence?" This would imply at least *some* recognition that human intelligence can't be reduced to a single ability. The Wikipedia entry for "intelligence" points to at least ten such abilities: the capacity for abstraction, logic, understanding, self-awareness, learning, emotional knowledge, reasoning, planning, creativity, critical thinking, and problem-solving. A 2007 study collected some seventy different definitions of "intelligence." Colloquially, we differentiate between book smarts and street smarts; intelligent quotients and emotional quotients (e.g., "he has a high EQ"). But turning to other species, we often fixate on only *one* kind of intelligence—the "human" kind (as if intelligence is readily reducible to one skill set, even across human beings and cultures).

Evolution emphasizes continuity among species. But it also highlights the beauty of difference. We humans inhabit one cognitive niche, but there are manifold others. Is one "better"? "To a lover of music," Nussbaum writes, "that's like asking whether we should prefer Mozart or Wagner: they are so different that it is a silly waste of time to compare them on a single scale."

Importantly, the point is not that humans aren't special. Under an evolutionary approach, every species is special! Humans undoubtedly excel at many feats, including cognitive ones. Human cumulative cultural evolution has enabled vast networks of cooperation and communication, and a remarkable ability to manipulate and control our environments. But this makes us neither more "evolved" nor superior to other species. For one, human intelligence involves all kinds of trade-offs—what the writer Justin Gregg describes as a "double-edged sword"—among them violent discrimination, raging pandemics, and a planet careening toward environmental collapse. If these don't make us think twice about our own wisdom, what will? What's more is that the ecological dominance of our species is often construed as *evidence* of our extraordinary intellect. But as we'll explore in a later chapter, this only conflates mastery with intelligence (it is interesting that even the word "mastery" itself has come to signify both comprehensive knowledge of something and control or superiority over something). *As if to dominate is to fully understand.*

To cut a long story short, there is no universally accepted definition of "intelligence," or way to measure it. Perhaps most generally, "intelligence" can be defined as how fast and successful organisms are in solving problems to survive in their natural and social environments. If chimpanzees had transformed the world but in doing so jeopardized their own livelihood, not to mention that of countless other species, would we be praising their intellect? In response to the ecological destruction caused by the dominant culture and economy, I've often heard people say we're just "too intelligent for our own good," but this view of intelligence remains stuck in seeing only one template—the human one.

Human Yardsticks

Members of *Homo sapiens* like to flaunt their superior intelligence, but other species would argue it all depends on the kind of tape measure you use.

A few years back, I had the strange opportunity to Skype with Kanzi, a bonobo who recently celebrated his fortieth birthday at a research center in Des Moines, Iowa. Throughout his life, Kanzi famously learned a lexigram system—a keyboard of abstract symbols— to communicate with humans using thousands of words. Kanzi is quite the celebrity in my field (a colleague, once prepping for an interview at the research center, perked up expectantly when an assistant entered the room with Starbucks coffee and snacks, only to realize that the delivery was for Kanzi instead). On the call, organized by a teaching fellow for my primate behavior course, the students and I observed as Kanzi pressed buttons on the lexigram to form novel sentences and (correctly) answer questions about various objects in his environment. For instance, when the experimenter pointed to an apple, Kanzi pressed the apple lexigram. Judging by my students' astonished reactions, it was easy to see why Kanzi had become a poster child for great ape intelligence.

Yet something didn't sit quite right. Much of my academic training had put ape language research on a pedestal—as key evidence for other species' intellect. But on that call, I couldn't help noticing how very *human* everything was: where Kanzi lived, the human experimenters around him, the English words they were teaching him. The surreal experience of Skyping with a bonobo.

A familiar theme emerged. It turns out that when we study the qualities that we think make us so smart, we are usually measuring the peculiarly human brand of these qualities (and as we've seen, usually not even a brand that's universal across the human species). Kanzi—like Koko, Washoe, and many other language-trained apes—is part of a long tradition of studies famously showing that captive apes can learn and use lexigrams or human sign language. Such studies may be interesting in their own right. But they are based on the implicit assumption that human language is the only possible language. They don't tell us much about the complex linguistic and other communicative abilities that other species use in their daily lives. A great quote attributed to the late astronomer Carl Sagan exemplifies the point: "It is of interest to note that while some dolphins are reported to have learned English—up to fifty words used in correct context—no human being has been reported to have learned dolphinese."

Sagan's message holds true today: a number of other species have learned to use human language far better than the reverse. But regardless of whether other species can learn aspects of human language, might they have their own languages that are just as fascinating? Consider the dance language of honey bees.

Karl von Frisch (the renowned ethologist we've met before) once observed that forager bees, upon locating food, will return to the hive and start performing an excited, figure-eight pattern dance across the hexagon honeycombs. Before long, the forager will recruit a number of followers, who retrace her steps before going out and locating the same food source (the phrase "make a beeline for it" likely derives from this behavior). Initially, von Frisch thought the bees must be picking up on some floral scent from the original forager that enabled them to locate the food. But years of painstaking observation revealed something much more interesting. The honey bees were encoding the precise direction and distance of the food source for other colony members. The duration of the dance is directly proportional to the distance of the food source: one second of dancing equates to roughly one thousand meters. The dance's angle, relative to the hive's vertical orientation, corresponds to the outward journey to the food source, relative to the sun's position. So a six-second dance at an upward angle of 30 degrees indicates food availability six thousand meters away at 30 degrees to the right of the sun.

The honey bee dance language is one of countless examples of a revolutionary scientific discovery made possible by looking past an anthropocentric lens. In truth, other species' languages should correspond to their umwelten, involving scents, colors, movements, gestures, patterns on their skin, and more. After all, it's called body language for a reason, and lizards may know it best. Sagebrush lizards express themselves in at least four ways: their posture, the number of legs on the ground, head bobs, and neck displays. Lest this seem simple, there are 6,864 possible combinations, 172 of which are frequently used. The skin patterns of Caribbean reef squid have also been likened to a language, potentially including grammar. Through rapid changes in skin color and texture, they communicate different messages on opposite sides of their body simultaneously (for instance, while a squid might threaten a male competitor on his right side, he might curry favor with a potential mate on his left). Do these majestic multitaskers look down on those of us who can't say two different things at once?

So often the knee-jerk reaction is to deny these feats as language—to insist that human language is something special. This is fair, so long as we extend the same consideration to other species. As the philosopher Eva Meijer clarifies, "Human language is special, but so is squid language, or bee language." When we define language by our standards alone, of course other species come up short. Isn't it more interesting—and indeed more scientific—to explore how studying these other expressions as languages can deepen and broaden our understanding of what language is and how it actually works?

Lying in my hammock, I watch the dance between shadow and light as sunbeams stream through the leaves of the maple tree overhead (the Japanese have a word—*komorebi*—for this very spectacle). From here I have a skylit window onto the birds' world, a glimpse into the avian umwelt. The other day, starlings were singing to one another at different layers of the tree's crown. It occurred to me how their arboreal, aerodynamic lives might shape their communication; how, unlike in humans, a typical conversation occurs across both horizontal and vertical planes. And again, I started to wonder. What if their call's meaning depends on the relative height between sender and recipient in the canopy? What if the distance between the interlocutors also matters? What if the position of the sun, the temperature, or the wind direction further alters the meaning of what they say? I'm not sure whether these questions have ever been asked, much less answered. My main point is that a world of possibilities—of what-ifs—opens when we let go of the idea that human language is the only true language. And in my experience, this is when science becomes a recipe for humility, and in turn for a rich and meaningful life.

⸻

Even to those who accept that the often maligned bird brain is capable of language, the idea that birdsong is *music*, a form of artistic expression, seems more dubious. But might the same arguments extend to other forms of expression?

Art is commonly considered the special province of humankind. According to one anthropologist, other animals do not have art, because their behavior "is largely determined by genetically controlled mechanisms, so that their responses are more or less automatic." Or as one psychologist wrote, "Music is characterized by an almost infinite variety, creativity, and novelty. The vocal signals observed in birds . . . in contrast, are hard-wired and have limited variability." And yet, as zoomusicologist Hollis Taylor argues, nearly half the world's approximately ten thousand bird species are songbirds, whose distinguishing feature is that they learn their songs! Birds are far from "automata" in their vocal development and production. Mockingbirds, for instance, have extraordinarily variegated songs with enormous repertoires, including most famously their ability to mimic other sounds within their environment.

There are also functional arguments, such as psychiatrist Iain McGilchrist's contention that "everything about human music suggests that its nature is sharing, non-competitive," whereas birdsong is "individualistic in intention and competitive in nature." But human music is not purely cooperative (just ask people trying to make a living in the cutthroat music industry). Nor is birdsong purely competitive. Songbirds regularly sing outside of territorial and other "competitive" contexts (plus, while a male bird serenading a female is a textbook example of male-male competition, isn't it *also* an example of cooperation with his love interest?).

Another argument is that birdsong lacks certain features characteristic of music—such as hierarchical structure, harmony, or large-scale coordination. But as Taylor notes: "One could propose a turnabout list of why human music does not reach the standards of birdsong because in Western art music the timbre of the individual participants is limited, the rhythm is pedantic, and improvisation is lacking." Taylor also notes how when we speak of music's defining "features," we're often fixated on the Western canon of music, highlighting the inextricable link between human exceptionalism and white supremacy. For the Kaluli people of Papua New Guinea, for instance, no sounds are performed in unison; instead, their music involves extensive overlapping, alternation, and layering. It's not any less sophisticated, just wonderfully different.

An amusingly enlightening film, *The Superior Human?*, runs through these and other highly referenced characteristics for human uniqueness, providing evidence for each trait in other species. I was surprised that "building" was on the list. Some have suggested that humans are unique in their ability to build complex structures. But what about the intricate constructions made by animals such as beavers, birds, and wasps? They might not look like our structures, but they are no less impressive. The world's largest beaver dam is nearly 850 meters long (it's visible from space, which is how scientists first discovered it). And, as will become a recurring theme, unlike many human-made constructions, beaver dams tend to promote biodiversity and ecosystem function.

You might say "But our structures are so much more immense— just look at our skyscrapers." Yet compared with the vast expanse of tropical reefs built by coral, which can stretch for hundreds or even thousands of miles (and are among the most diverse ecosystems on earth), human structures do not seem uniquely grand. Plus it's all relative. The tallest human-made building in the world, the Burj Khalifa in Dubai, is 828 meters tall—531 times taller than the average human. Termites can construct mounds up to 9 meters, about 914 times taller than the average termite. In other words, this would be like humans building a 1,424-meter-high building!

Termite mounds don't get enough credit, so let me elaborate. The first time I ever saw one, in Kenya, I admittedly thought it was a very large pile of sand—an outsize drip castle, like the ones we used to make as children on the beach. But termites' mounds are remarkably intricate structures, with porous walls for cooling and heating the mound, elaborate features for draining water, and even areas where they cultivate fungi as a source of food and medicine. We like to think that humans invented agriculture, but termites had it figured out long before we did. Their cities can support millions, and they've been building them for ages. Architects around the world today are modeling their constructions after termites' design methods. Take the Eastgate Centre in Zimbabwe, the country's largest office and shopping complex. Inspired by the ventilation system of termite mounds, this building has no conventional air-conditioning or heating yet maintains year-round climate control (with dramatically less energy consumption). How much do we have to learn from innovations successfully honed over millions of years of experimentation—a process otherwise known as evolution by natural selection?

Nature's Greatest Plagiarizer

The term "biomimicry" refers to human efforts to design and produce structures and systems that imitate Nature, and today there are countless examples.

Did you know that humpback whales' fins inspired modern wind turbine blades? Colossal creatures like whales move through the water so elegantly in part thanks to bumps on their fins, called tubercles. Tubercles decrease drag and increase maximum lift, a feature scientists have co-opted to boost wind turbine efficiency.

Bullet trains like the Shinkansen in Japan are modeled after kingfisher birds, who deftly move between air and water without making a splash. Eiji Nakatsu, the Shinkansen's chief engineer and an avid bird-watcher, designed the train's nose after a kingfisher's beak, making it run faster with significantly less noise and energy consumption.

The textile industry also regularly emulates Nature's wisdom. Velcro was invented by the Swiss engineer George de Mestral in the 1940s after he noticed and tried to remove the burrs (from a burdock plant) sticking to his Irish pointer's hair. Meanwhile, deep-sea sponges can construct intricate glass fibers that are stronger and more efficient than anything humans can yet make. The first commercially practical glass fibers were not invented until the 1930s, but as Harvard materials scientist Joanna Aizenberg points out, "Sponges knew how to do it a half-billion years ago." In similar fashion, by mimicking how a spider spins silk, companies like Spintex—whose clever tagline reads, "Backed by 300 million years of research and development"—are creating high-performance, sustainable textiles that are one thousand times more efficient than an equivalent synthetic fiber.

Nearly every day in the arid Namib desert, I would come across jet-black, long-legged beetles engaging in a peculiar behavior I could describe only as an attempted yoga headstand (a very scientific estimation indeed). Only later did I learn that they were doing something much more remarkable—literally pulling water out of thin air. Leaning their bodies into the wind, Namib desert beetles allow fog droplets to accumulate on their backs, which eventually get rolled toward their mouths. Inspired by these beetles, human researchers are now developing biomimetic technologies capable of harvesting water from the air to help solve water-shortage problems.

Art is perhaps the area where Nature's influence is most evident. Birdsong has inspired musicians from Vivaldi to Pink Floyd. Miles

Davis apparently honed his sound by going into the woods to listen to birdsong. The composer Olivier Messiaen went so far as to state, "Birds are the greatest musicians on the planet." Would Monet's water lilies and O'Keeffe's orchids and Van Gogh's sunflowers have ever been possible without more-than-human conspirators? And this is to say nothing of written works. "Nature," Thoreau famously penned, "is full of genius."

On a recent visit to Amsterdam's Stedelijk Museum, I strolled through the installation *MycoTunnel* by the Ghanaian Filipina artist and architectural scientist Mae-ling Lokko. The tunnel's clean, modern feel belied its organic makeup: rows of mycelium (the root structure of fungi) fed by agricultural waste. More and more artists and scientists are using mycelium and other bio-based materials to mimic the circular designs and nutrient-cycling processes found in Nature (for which fungi are especially known). As with other biomimetic structures, these materials not only tend to outperform others on the market but are also much more sustainable from start to finish. Walking through *MycoTunnel*, one can't help appreciating a future world in which we embrace biomimicry on an even greater scale. Other species have shaped their environments for eons with ingenuity and resourcefulness—who could think of better allies as we navigate the present ecological circumstances?

One afternoon, I sat with my students under an old yew tree in the courtyard near our classroom. One of them was rubbing some yew berries between his fingers, staining his notebook a deep red hue. "Wait a second," he said, before asking the obvious question: "Sure, other species have all these amazing abilities, but haven't humans developed technologies to make up for what they lack?" Yes, birds can fly, but humans created airplanes (so one version of the argument goes); isn't this evidence of our superior intelligence as a species? "Fair question," I replied, before attempting to turn the question around. "But couldn't our technologies *also* be a weakness, a sign of human vulnerability and dependency?" (Coincidentally, university Wi-Fi was malfunctioning that week, causing major disruptions around campus, a firm reminder of how overreliant on it we've become.) Another student chimed in: "Plus there are trade-offs to our technologies"—the double-edged sword idea. "Just look at our relationship to social media!" But I could tell my student wasn't convinced. After a few minutes of back-and-forth about the pros and cons of technology, another classmate took a different tact: "Well, other species usually had things figured out long before we did, so is it really fair to call it *human* innovation—to celebrate everything as some great feat of *our* imagination?" Following the airplane example, the Wright brothers famously devoted hours to studying birds' movements in the wind. According to local citizens near their home in Kitty Hawk, North Carolina: "They'd stand on the beach for hours at a time just looking at the gulls flying, soaring, dipping." They were especially intrigued by gannets, the seabirds with giant (up to six feet) wingspans. "They would watch the gannets and imitate the movements of their wings with their arms and hands. They could imitate every movement of the wings of those gannets." Would humans have ever conceived of airplanes were it not for some kind of avian inspiration? "Learning the secret of flight from a bird," Orville once said, "was a good deal like learning the secret of magic from a magician."

My partner, who works in health care, considers evolution the supreme medicinal innovator. We often talk about biomimicry in modern medicine, most recently the revolutionary "discovery" of CRISP-R, a gene-editing technology that was actually discovered by bacteria (specifically, the bacterial immune response to a virus). It's not only sobering to consider all the medicinal possibilities we'll never realize when forests and other ecosystems are destroyed (nearly half of all Western medicine comes from plants); it's also humbling to wonder whether great feats of "human" imagination like CRISP-R would have even been possible without other species getting there first. When we chalk technologies up to human ingenuity alone, as my skeptical student later pondered, isn't it like claiming someone else's ideas or work as our own? "When we don't give credit where credit's due," he wondered aloud that day while eyeing the university honor code on the syllabus, "aren't we plagiarizing Nature?" The implications of his question make a world of difference for how we honor and appreciate other species and their role in our lives.

Of the First Rank

Intelligences are many and varied, but there is still a tendency to consider primates the highest-ranking of all animals, and to consider humans primates par excellence. We are so convinced of this primacy that it's reflected in the name Carl Linnaeus gave to our taxonomic order: Primata, meaning "of the first rank."

I am proud to be a primatologist. I feel fortunate every day that *this* is what I get to do—to think and work with other species—a childhood dream come to life. But sometimes even *I* worry that the disproportionate attention paid to our close primate relatives is just another reflection of anthropocentrism. Chimpanzees and bonobos regularly make headlines for challenging the supposed human-animal cognitive divide. But is this all that surprising? We should expect (and indeed do find) that closely related species tend to have similar forms of cognition. This does not make primates more clever than other animals, just clever in ways with which we—human primates—can readily identify (and, I might add, can thus conceive of more species-appropriate studies to test!).

My friend the ethologist Marc Bekoff called this pervasive tendency to prematurely dismiss the cognitive skills of non-primates "primatocentrism." The comparative psychologist Benjamin Beck dubbed it "chimpocentrism," emphasizing the overwhelming focus on great ape intelligence. Admittedly I, too, have fallen into the trap of seeing primates as especially intelligent, simply because their intelligence resembled our own—it was one of the reasons that drove me to study them in the first place. Unlearning my own human-centric biases and becoming aware of their hidden influence is never-ending. Case in point: I have used the word "primate" countless times in my career and had not, until writing this book, once considered its deeper connotation.

I often see papers highlighting human achievements and identifying *forerunners* of the relevant capacities in our primate relatives— the "building blocks" of human cognition: Chimpanzees can adopt others' perspectives, but only humans have full-fledged theory of mind. Baboons can recognize numbers, but only humans do math. Other primates have "pre-culture," not culture; "proto-language," not language. If other primates are cooperative, then we humans are "hyper-cooperative"; if they are social, then we are "ultra-social." As Thomas Suddendorf describes in a 2013 CNN interview, we have transformed "common animal traits into distinctly human ones— communication into open-ended language, memory into strategic planning and social traditions into cumulative culture." Studying our closest living relatives can help us understand the evolutionary roots of human abilities—an interesting and often humbling endeavor in its own right. But as these examples show, sometimes the quest portrays other species narrowly, as less complex versions of human beings—with rudimentary, not-quite-human minds. As if *their* minds have remained unchanged for eons while *ours* continuously evolved. As if all animals are on their way to becoming human; they just didn't get there yet.

That same general bias is also evident when people equate other animals' mental capacities with those of human children. This includes popular claims that a chimpanzee's (or a dog's) intelligence is on par with that of a two-year-old child. While this may be a useful comparative lens for understanding the development of certain human cognitive skills (such as the ability to understand human words or gestures), it overlooks that other animals have their own uniquely evolved capacities (if you dropped a human toddler and an adult chimpanzee in a forest together, who do you think would appear more intelligent?). Other species' minds are not ancillary or intermediary to ours. "All living beings are the tips of the so-called tree of life," explains the philosopher Colin Allen. "Thus there is no evolutionary reason to expect that rabbits or chimpanzees should lie somewhere midway between humans and any other organism on any particular scale, whether it is of intelligence or ear shape."

"So Like Us"

Conservation research is often biased toward protecting those species most similar to us. Even more formally, having human-like characteristics has become a moral dividing line. In courtrooms worldwide today, we determine how we treat other beings based on the extent to which they have minds with features like our own—what Martha Nussbaum termed the "so like us" approach.

Some animals—a select group that includes great apes but more recently also elephants and whales—are thought to be enough "like us" to deserve moral respect and legal protection. Advocates of this approach, in particular legal thinkers Steven Wise and the Nonhuman Rights Project, are now representing individual animals in courts (perhaps most famously, Tommy, a chimpanzee kept in upstate New York; and Happy, an Asian elephant kept at the Bronx Zoo). The project is filing habeas corpus writs, traditionally meant to protect bodily liberty and contest human prisoners' illegal confinement. A court that granted the writ to other animals would be acknowledging their legal "personhood," their right to be free. This would have an enormous impact on modern institutions involving intensive animal confinement, including labs, zoos, and factory farms.

At first glance, by aiming to extend protections that have been conventionally reserved for humankind to other species, these cases seem to counter the dominant culture of human exceptionalism. But various scholars say the general approach risks doing just the opposite. Rather than challenging human exceptionalism, they argue, the "so like us" approach plays upon the familiar idea of the *scala naturae*, which puts humans atop a hierarchy and ranks other species according to their likeness to humans. Nussbaum clarifies her reasoning here:

"In short, if we line up the abilities fairly, not prejudging in favor of the things we happen to be good at, many other animals 'win' many different ratings games. But by this time the whole idea of the ratings game is likely to seem a bit silly and artificial. What seems truly interesting is to study the sheer differentness and distinctiveness of each form of life. Anthropocentrism is a phony sort of arrogance. How great we are! If only all creatures were like us, well, some are, a little bit. Rather than unsettling our thinking in a way that might truly lead to a revolutionary embrace of animal lives, Wise just keeps the old thinking and the old line in place, and simply shifts several species to the other side."

Does the "so like us" agenda perpetuate human exceptionalism by continuing to make humans and human abilities the measure of all things (and in doing so, reinforce the very model of the world it apparently challenges)? Or is it a necessary and logical first step, short of completely overhauling our present legal system? What, if any, are the alternatives? Hold these questions for now, as we'll return to them later.

———

Whether in zoos, labs, sanctuaries, or the wild, other primates have changed my understanding of how life and evolution work. The more

I've studied and learned with them, the more the supposed human-animal divide falters, and my preconceived notions about human uniqueness fade. An ever-growing body of research shows how our close primate relatives are "like us"—in their language abilities, social inclinations, conflict resolution strategies, emotions, and more. My doctoral research ultimately showed that human and chimpanzee reconciliation shares similar underlying mechanisms—namely a broader tendency to switch between different motivational states. But does this make them more human-like, or us more chimpanzee-like? How do we know which way the arrow points?

In my view, the differences in how we go about reconciling are just as interesting. By some scales, our strategies could be seen as more effective and sophisticated. But by other metrics, theirs could be. For instance, humans often have lengthy discussions about areas of conflicts in an attempt to resolve them. *But language is no guarantee of understanding.* How many times have you been in a conflict with someone where "talking things out" actually made the disagreement worse, or became the new conflict at hand? Other primates engage in different kinds of affiliative social interactions (like grooming and embracing) following conflicts, and often swiftly resume regular activities together. I once witnessed two adult males after a particularly heated conflict quickly embrace and start walking in sync, literally and figuratively moving on. This inspired my colleagues and me to write a theory paper about how going for a walk together could promote conflict resolution between humans, rather than lengthy sit-down conversations, which sometimes leave partners further entrenched in their conflict. Research has since tested and confirmed our general predictions. Understanding how we are "like chimpanzees" helped us understand human ways better. It's not that chimpanzees' reconciliation strategies are a more rudimentary form of our own. Just similar in some fundamental ways, and different in others.

As journalist Philip Ball recently wrote for *The Guardian*: "The challenge, then, becomes finding a way of thinking about animal minds that doesn't simply view them as like the human mind with the dials turned down: less intelligent, less conscious, more or less distant from the pinnacle of mentation we represent. We must recognise that mind is not a single thing that beings have more or less of." As an alternative, Ball draws on a concept first proposed in 1984 by the computer scientist Aaron Sloman: the "space of possible minds." According to this idea, there are many possible dimensions of mind. And as Ball notes: "We are no more at the centre of this mind-space than we are at the centre of the cosmos. So what else is out there?"

Erwin's Law

What else *is* out there? The truth is, we still have a lot to learn, in part because scientists have conventionally gravitated toward studying the species that most resemble our own. So much remains unknown about the vast majority of beings out there. This is also known as Erwin's Law, coined by the applied ecologist Rob Dunn—which states that life is far less studied than we imagine it to be. The law is named for the entomologist Terry Erwin, who revolutionized our understanding of the dimensions of life on earth with a single study in a Panamanian rainforest in the 1980s. That study led to Erwin's estimation that there may be up to thirty million arthropod species living on the planet (previous estimates put that number around 1.5 million). It was a humbling insight, one that magnified the limits of our scientific knowledge. As Dunn highlights, perhaps seven out of eight animal species remain unknown to science. And from what we can gather today, the "average" animal is neither like us nor dependent on us. They are not a vertebrate nor even a mammal; instead, they're a tropical beetle, moth, wasp, or fly.

And it doesn't end there. Our perspective is not just primate- and human-centric but also zoo-centric. On average, we are probably more informed about animal life than we are about every other species out there. In a photograph of an environment composed of 90 percent vegetation, the presence of a single animal will occupy participants' attention within the entire scene—a "plant blindness" that makes us underappreciate the flora around us. The article coining the term alerted scientists to the bias of valuing animals over plants "because biology is taught by members of the animal kingdom." Trees, flowers, and shrubs are frequently lumped into single categories that mask a staggering amount of evolutionary diversity. While the English dictionary has only a single word for them, there are more than sixteen thousand known species of moss, and more waiting to be discovered.

Fungal blindness may be even more common than plant blindness. Yet fungi are more closely related to us than they are to plants. Though we interact with them constantly (with every inhale, we breathe in between one and ten fungal spores), we know very little about them. In a study of the fungi found in houses across North America, Dunn and his colleagues found that no fewer than half of those species may be new to science. And what about bacteria? Some estimate that there are a trillion (that's 1,000,000,000,000) kinds of bacteria on earth. As Dunn writes: "Erwin's estimate had led scientists to imagine that most species were insects. For a while, it seemed as though fungi might be the big story. Now it seems as though, to a first approximation, every species on earth is a bacterial species." Billions of these bacterial species await discovery, some of them dwelling in and on your body at this very moment. "Our perception of the world keeps changing; more specifically, our measure of the dimensions of the biological world keeps expanding," notes Dunn. "And as it does, the average way of living in the world seems to be less and less like our own."

Erwin's Law beckons us to have more humility. Organisms not yet studied are likely very different from those who have been, meaning we understand very little about how life on earth works. Anthropo-centric biases instead give us a false impression of the world—of how species behave, look, think, and evolve. After all, as my students often joke: there is one evolutionary biology department at Harvard dedi-cated to studying humans, and another, *to the rest of life on earth*!

An interest in human evolution is understandable—after all, we are humans. And the human umwelt certainly renders it challenging to really understand those species who are *unlike* us. We may come to appreciate their intelligence, but their experiences may always remain practically unknowable to us because of the radical difference that exists between our lives. Given our strictly human point of view, the barrier is insurmountable. Or is it?

CHAPTER SIX

Thinking Otherwise

W HAT IS IT LIKE TO BE A BAT?" PONDERED THE PHILOSO-
pher Thomas Nagel in a classic 1974 essay of the same title. In what might be the most widely cited paper in the field of the philosophy of mind, Nagel wrote that the conscious experiences of other animals are inherently subjective and hard to describe. We can form an impression of their umwelten—their sensory worlds—by understanding their behavior and perceptual abilities, but we'll never know what it's actually like to *be* them.

One could try to imagine, Nagel wrote, "that one has very poor vision, and perceives the surrounding world by a system of reflected high-frequency sound signals," or that "one has webbing on one's arms, which enables one to fly around at dusk and dawn catching insects in one's mouth." But that wouldn't cut it. This would tell you only what it would be like for *you* to behave as a bat behaves. Instead, Nagel clarified, "I want to know what it is like for a *bat* to be a bat. Yet if I try to imagine this, I am restricted to the resources of my own mind, and those resources are inadequate to the task." What it's like to be a bat, he ultimately concluded, is a question that we will likely never answer; it lies "beyond our ability to conceive."

Mind Problems

For as long as I can remember, I've been fascinated by questions about consciousness. Why did consciousness evolve? How does it correspond to the physical world? What gives rise to this "me" dwelling within and through my body—to my subjective sense of self? And what is consciousness like in other lifeforms? Sometimes I lie awake pondering these questions until existence becomes viscerally strange. Ironically, it's the moments before losing consciousness to sleep that I occasionally feel closest to answers. Nearing an epiphany, I suddenly lose the train of thought completely (my mind on rapid rewind, catching bits and pieces but ultimately left with nothing coherent to grasp)! I'm right back where I began. As if the universe isn't quite ready for me to know. As if the universe knows *I'm* not quite ready to know.

Scientists and philosophers have grappled with similar questions for millennia. "There is nothing that we know more intimately than conscious experience," wrote the philosopher David Chalmers in a landmark essay on the topic, "but there is nothing that is harder to explain." In that same essay, Chalmers introduced his influential distinction between the "easy problem" and the "hard problem" of consciousness. The easy problem is understanding how physical phenomena (such as brain processes) give rise to cognitive functions like learning, perception, and action. The hard problem is understanding how those processes give rise to *experience* (i.e., phenomenal consciousness or mental states with phenomenal qualities, or *qualia*). The bitterness of grapefruit, the softness of velvet, the blueness of the sky—some scholars admit they have no idea how to account for experiences like these, and suspect they never will.

The easy problem is generally thought to be solvable by the traditional methods of science. But the hard problem is widely considered the central challenge faced by consciousness research today. Though recent advances in neuroscience are encouraging, it's unclear whether *any* amount of scientific progress will be able to solve the hard problem of consciousness. What gives rise to conscious experience might always remain a mystery. But importantly, that mystery applies to bats *and* humans alike. In other words, there isn't just a hard problem when it comes to other species, but a hard problem in general.

The definition of "consciousness" most philosophers use today—one frequently traced back to Nagel's essay on bats—is that an entity is conscious if there is *something it is like to be* that entity. In Nagel's example, a bat is conscious because there is something it's like to be one, some subjective experience there, though we humans can't experience such a mode of consciousness directly. That same general objection, that "we can never really know what they experience" is still frequently heard in debates about animal consciousness today. Because we lack direct access to others' inner states, we can only make inferences based on what we can observe (in philosophy, this is another problem, known as the "problem of other minds"). But to the extent that this is true, might it also be true of *all* minds—including, dear reader, of yours and mine?

The Third Wound

During fieldwork, you spend hours on end observing animals alongside colleagues, often chatting to fill the silence. But despite all the talking, the close proximity and remote setting can lead to miscommunication and tensions among team members, an underappreciated challenge of fieldwork that I've experienced firsthand. You might find yourself more attuned to the inner worlds of the animals whose behaviors you've carefully been attending to than to those of the humans you've been conversing and working with for months.

It is commonly believed that inner states are harder to demonstrate in other species because (unlike most humans) they cannot verbally report them. Yet while language can undoubtedly help, our ability to use words to relate feelings is not all it's cracked up to be. If I told you that I was feeling stressed, you'd have some insight into my inner world. But does this give you direct access to my experience? Can you know with absolute certainty what I feel? Is my subjective experience of "stress" the same as yours?

As much as language can facilitate understanding, it can also easily deceive, distract, and derail. Many humans are notoriously bad at understanding and communicating what they actually *think* and *feel*. My aunt has turned this challenge into a career. As a clinical psychologist, she gets paid to help people get more in touch with their own thoughts and emotions.

This is one of the reasons self-report methods have come under more scrutiny in psychology research. We misrepresent our own private experiences for all sorts of reasons. We may be unaware, or we might deliberately stretch the truth or respond in ways we deem socially desirable. One study led by Robert Feldman found that most WEIRD adult participants lied at least once during a ten-minute conversation, especially when told by experimenters to appear likable or competent to their conversation partner. Humans can be frighteningly good at self-delusion. "The conventional view that natural selection favors nervous systems which produce ever more accurate images of the world must be a very naïve view of mental evolution," wrote the evolutionary biologist Robert Trivers, who spent much of his career trying to understand how this capacity for self-deception came about. His theory, for which there is compelling evidence, is that we lie to ourselves because it makes it easier to lie to others.

On a scale of 1 to 7, please indicate how well each of the following statements describes your thoughts and feelings. During my dissertation research, I regularly participated in psychology studies on my lunch break. I mostly wanted to learn how to develop my own survey methods to study human subjects. On the surface, it seemed like a luxury to be able to ask participants about their motives. But having worked exclusively with other primates up to that point, I remember feeling skeptical. Can we trust what human primates say about themselves?

I immediately help those who are in need. Circling my answers, I couldn't help wondering: Are those who consider themselves helpful the most inclined to help? *I easily share with friends any good opportunity that comes to me.* Instead of asking people whether they share, why not just study their behavior—what they actually do? These surveys made me appreciate the value of behavioral observations even more. *I try to console those who are sad.* I sensed that my answers to these seemingly straightforward questions varied depending on myriad factors—my mood, the weather, the people I had recently interacted with. And as I occasionally learned after being debriefed about the study's goals, they also depended on many factors of which I was wholly unaware. So-called subliminal stimuli—those occurring below the threshold of conscious awareness—can shape everything from self-esteem to our willingness to help and empathize with others.

The zoologist and psychologist Sara Shettleworth famously argued that human behavior expresses unconscious responses to simple cues more often than we like to admit. It's the basis of the third "narcissistic wound" to the human ego, added by the founder of psychoanalysis himself, Sigmund Freud: first it was Copernicus (humans do not live in a geocentric universe), next it was Darwin (humans are animals), and then it was Freud (humans are not the masters of their unconscious).

Freudian ideas have a controversial reputation, but there is no denying the role of unconscious processes in controlling our behavior. We're gliding around on autopilot much of the time—current scientific estimates indicate that some 95 percent of human brain activity is unconscious. Our communication is influenced by the workings of pheromones and other unintentional molecular processes (we say people have "good chemistry" for a reason). Most perception occurs outside our awareness, which is why subliminal advertising is effectively banned in most countries. The unconscious mind plays a much deeper role in our mental operations—including those involved in memory, perception, reason, and even language—than we often acknowledge.

Skeptics about other species' mental abilities sometimes assume that the corresponding human ability relies on conscious experience (and thus that conscious experience is an essential feature of the ability in question). To illustrate, consider doubts about whether other animals possess episodic memory. These doubts often hinge on the tacit assumption that the memories humans report derive from a conscious replay of past events. Yet human memory is less reliable and dependent on conscious recall than we think, creating notorious problems with eyewitness testimony in court cases. The tendency to define certain psychological processes by an exaggerated account of typical human performance in order to deny them to other species even has a name (and it's a mouthful): anthropofabulation.

Heavy Burdens of Proof

Let's now revisit the common objection that we can never really know what other species experience—or in Nagel's terms, *what it's like to be* them. Just yesterday, one of my students raised this very concern. On her way to class, she came across two brown rats pouncing playfully on each other in a small patch of grass. Judging by their spirited hops and reunions, they were having fun. But, she questioned, how can we know the rats were experiencing joy? How can we be *sure*?

The truth is, we may never be 100 percent sure. And so it is with all subjective experiences. In humans we routinely rely on various forms of imperfect and indirect evidence (including how other people behave and what they say about how they feel) to infer the existence of what we cannot directly observe. Some of that evidence (e.g., expressing feelings in words, an indirect source of information) may not be available to us in quite the same way it is in other species, but there are so many other sources of evidence!

Decades of research have shown that body language and other nonverbal cues are just as important in communication as the spoken word. We often use these other sources of information to understand people whose languages we do not speak. For instance, my partner, Lucas, and his family are from the Netherlands. His young nieces and nephews have only begun to learn English in school, and my Dutch is still shaky. But just because we don't speak the same language doesn't mean we don't understand each other. We have to get a bit more creative—attending to facial expressions, gestures, and the like—but it sometimes feels like an even deeper form of understanding can result. As the Dutch philosopher René ten Bos once said: "Jij en ik . . . juist door te praten een hoop dingen niet [laten] zien" (you and I . . . precisely by talking [leave] many things unseen).

Just imagine being on a boat rocking back and forth as the ocean swells intensify. You go out to the deck for some fresh air and see someone—a total stranger—pale, sweating, and retching. There is good reason to believe that they are seasick, even if they assure you to the contrary. You can't know how they feel for sure (with absolute certainty), but you can be reasonably sure. Likewise, when a dog yelps after stepping on a wasp, limps, and protects his injured foot, we assume that he experiences something more than a simple, painless reflex. We assume that the human and the dog are feeling something bad. We are using probability and common sense.

Of course, there is always the risk that we might misread the behavior of the individual in question and incorrectly interpret their experience. There are many mysteries in animal (including human) behavior and being; it's part of the beauty and trouble of living together. But most of the time, it works pretty well. And surely it's better than assuming we can infer nothing about the private feelings of those around us. Yet we seem to require more certainty when it comes to other species. We tend to emphasize how we can never *really* know whether a particular being is conscious (note this means it's just as impossible to prove that they are *not* conscious as to prove that they are). That demand for perfection of evidence appears to be a subtle way of adhering to human exceptionalism—a contemporary myth often cloaked in scientific rigor.

According to the founder of cognitive ethology, Donald Griffin, "The tendency to demand absolute certainty before accepting any evidence about mental experiences of animals reflects a sort of double standard." The requirement for perfect evidence, he points out, would have seriously impeded progress in almost every area of science to date. Twentieth-century biologists studied chromosomes even though they couldn't yet see the precise nature of their genes. Dark matter has never been observed directly, but without it, our theories of gravity can't explain astrophysical phenomena. Science deals with many other unobservables, like the Big Bang and continental drift. Consciousness is another one of those unobservable phenomena, at least for now.

It was only in the second half of the twentieth century that this "commonsense" view of human consciousness began to gain traction in science (recall that up until then, behaviorism had effectively prevented the study of human as well as animal consciousness). Once we allowed ourselves a peek under the behaviorist carpet, there was a huge increase in research on human consciousness. We assume that human participants are conscious agents every day, even though we lack direct access to their conscious states. How might we do science differently if we extended this same assumption to other species, so that our task is to understand how and what (not *whether*) they think and feel?

As we've learned, even if other species don't exhibit intelligence like ours, we can't conclude that they are not completely or meaning-fully intelligent—different does not automatically equal less! Can the same be said for other dimensions of mind—including conscious-ness? Under an evolutionary perspective, cognitive abilities arise to solve for particular functions. Yet consciousness historically eludes definition in functional terms. After all, as my students occasionally gripe, the most precise definition of consciousness is "something that it's like to be." But this "something" is no trivial matter. Evidence of consciousness typically carries more moral weight than evidence of intellect—in part because consciousness underpins the capacity to suffer.

"But 'Can They Suffer?'"

In 1789, Jeremy Bentham, the founder of utilitarianism, famously stated that the critical question regarding the moral status of animals is not "'Can they reason?' nor, 'Can they talk?' but, 'Can they suf-fer?'" In other words, the foundational qualifier for moral considera-tion is the ability to consciously *experience* feelings and sensations (a.k.a. sentience) that enable states like pain and pleasure.

There is growing appreciation among scientists and philosophers that many animals are sentient. The Cambridge Declaration on Con-sciousness, published in 2012 by a prominent team of neuroscientists, claimed there was sufficient evidence for consciousness in all mam-mals and birds and even cephalopods like octopuses. But there are fierce, ongoing debates about whether certain creatures including fish, crustaceans, and insects experience conscious states like pain. And as in debates about intelligence, human exceptionalism often looms large.

The usual argument against other species feeling pain—or for that matter, having consciousness at all—is that their brain structure is so different from ours. The reasoning goes something like this: we know the areas involved in pain in humans; if you lack those areas, you must not feel pain. But there is a fundamental problem with this line of thinking. Evolution tells us time and again that different structures can serve similar functions. Crustaceans don't have a visual cortex anything like that of a human, but they can still see. Most fish don't have lungs like ours, but they still manage to breathe (likewise for all insects and plants!). No good physiologist would consider the human digestive or immune system a fitting reference point for describing the variety and complexity of animal metabolism or immunity (platypuses digest their meals just fine without a stomach). So why when it comes to the *mind*—whether it's cognitive capabilities like intelligence or conscious states like pain—do we continue to make man the measure of all things?

Of course, denying (or conveniently doubting) other species' ability to feel pain helps us manage dissonance about harming them. Again, this is ironic given the prominent role of animal experimentation in human pain research, which presupposes that we have similar biological and psychological responses to pain. But skeptics about other animals' experiences of pain usually provide other justifications.

At the core of these debates is the distinction between nociception and pain. Nociception is the ability to detect noxious stimuli, usually accompanied by a reflex withdrawal response. Pain is the conscious experience of suffering that ensues. Though nociceptive stimulation usually leads to pain, pharmacological and neurological research shows that one can exist without the other (not to mention personal experience, as when we accidentally touch a hot pan and jerk our hand back reflexively, before we feel any pain). From an evolutionary-function standpoint, nociception is an adaptive response that helps protect the body against tissue damage. All animals are considered capable of nociception. But for many species—especially those who lack a particular area of the brain called the neocortex—it's thought to end there: a reflex, with no "felt" pain. The inclusion of birds and octopuses in the Cambridge Declaration already challenged this conventional wisdom. Members of both taxa lack a mammalian neocortex but exhibit behaviors and other neurological patterns indicative of consciousness.

It's worth noting here that Donald Broom, a biologist specializing in animal welfare, once likened the nociception-pain distinction to "a relic of attempts to emphasize differences between humans and other animals or between 'higher' and 'lower' animals." After all, Broom notes, for other senses such as vision and hearing, the action of sensory receptors and the subjective experience aren't given different names. Scientists who study eyes don't usually debate whether humans have vision and other species like fishes and flies merely have photoreceptors.

Nonetheless, even if we *do* make such distinctions (as many scientists today do), more and more evidence suggests that the responses of these supposedly "lower" species are not reducible to reflexes. In a groundbreaking (though somewhat painful to read) study, the zoologist Lynne Sneddon and colleagues injected trouts' lips with toxins. Unlike those injected with saline, these poor trouts lost their appetite for several hours. They displayed anomalous behaviors including rocking back and forth, breathing heavily, and rubbing their injured lips against the side of the tank. Later studies found that such responses are ameliorated when scientists give the animals painkillers. Similar evidence exists in invertebrates like crustaceans. For instance, when acetic acid is brushed on one of their antennae, prawns will engage in prolonged rubbing and scraping of that specific antenna, a behavior that subsides when local anesthetic is applied. Moreover, an injured crab will guard his wound by putting a claw over it whenever a rival crab is around.

These behaviors, which persist well past the injury, are difficult to explain via nociception alone. They are all consistent with the idea of pain. We may not *know* with absolute certainty, but the evidence—as far as many of us are concerned—puts us beyond reasonable doubt. We're still not exactly sure why pain evolved and what its function is (why not just the reflex, which itself prevents further immediate damage?), but some speculate that the experience of pain helps us learn and remember to avoid those threatening situations in the future.

Many people, while acknowledging that animals experience pain, harbor an underlying assumption that human pain is somehow more intense. Even the concept of evolutionary continuity can evoke a spectrum of pain with humans as the end point. But even here one could argue just the opposite: there is reason to think that some species might feel even *more* pain than humans do. For many animals, humans among them, experiences of extreme pain often cause their nervous system to shut down as a coping mechanism. Unfortunately, lobsters and other crustaceans don't have this ability to go into "shock." So when they are exposed to cruel procedures (like being boiled alive), their suffering may be prolonged. Or consider squids. If a squid's fin gets injured, nociceptors are activated not only in the region of the wound but across a large part of her body, reaching as far as the opposite fin. She may hurt all over, rather than in one targeted area. These findings run counter to the idea that other animals experience pain less intensely than we do. As with intelligence, there may be diverse types of pain and conscious experience.

And as with debates surrounding intelligence, these debates

seem to move in only one direction, starting with species "like us" and eventually bringing more distant species into the fold. Beginning with the human template only to recognize that minds manifest in many different ways. Will insects be next? There is growing evidence to suggest they will be, and it becomes a more urgent ethical matter each day. Annually, *trillions* of insects are farmed, killed with pesticides, or used for research purposes—trends set to grow in the coming years. But insects are almost universally excluded from animal welfare legislation, at least in part thanks to the prevailing assumption that they do not feel pain—that their brains are too tiny, or lack the appropriate neural architecture, to support sentience.

"There may be extraordinary mental activity within an extremely small absolute mass of nervous matter," marveled Darwin, who stood in awe of the "wonderfully diversified instincts, mental powers, and affections of ants"—all despite their minuscule brains. "Under this point of view," he wrote, "the brain of an ant is one of the most marvelous atoms of matter in the world, perhaps more so than the brain of man." Recent research shows that, despite their modest appearances, insect nervous systems are exquisitely complex and may perform many of the same functions as mammalian nervous systems. A growing number of studies show animals such as bumblebees and fruit flies engaging in behaviors indicative of positive and negative emotional experiences—all in the absence of human-like brain structures.

There are still substantial gaps in our knowledge (recall Erwin's Law—we know hardly anything about the vast majority of insects out there). But even the *possibility* that insects are sentient has profound moral consequences (as philosopher Jeff Sebo points out, even a small but non-negligible chance that trillions of individuals are sentient adds up to a lot of potential suffering!). And while insect farming is often viewed as a sustainable and ethical alternative to conventional animal agriculture, it may not be a panacea, like its supporters claim. For one, insect farming and traditional animal farming are presently mutually reinforcing systems (many new insect farms are selling insect meal primarily to huge aquaculture operations, and lobbying to expand to chicken and pig factory farms). But on a deeper level, what set of relationships and beliefs about the world does the insect farming industry reinforce? Does touting the mass commodification of insects (over allegedly "higher" animals like cows) draw on and validate the familiar idea of the *scala naturae*? What would it mean to avoid the same false hierarchies we've constructed in the past?

When it comes to other minds, can we build on foundations of better evolutionary theory and less anthropocentric assumptions about brain function?

Brain Power

Earlier we explored the various ways psychologists have completed The Sentence ("The human being is the only animal that . . ."). Traditionally the focus was on behavioral and cognitive hypotheses like "uses tools," "has culture," "is rational," and so forth. But more and more today (in part thanks to the available technologies), the spotlight has turned to genetic and other biological markers of human uniqueness.

Yet when we look at our genomes, a similar picture emerges— genetically we are far more similar to other species than different. Nearly every gene in the human genome (98.7 percent, to be precise) has a counterpart in the genomes of our closest living relatives: chimpanzees and bonobos. We three species are more closely related to one another than we are to gorillas or any other primate. Perhaps even more counterintuitively, we humans don't necessarily have larger genomes than so-called "lower" taxa. Our kind has fewer genes than many plant species and fewer base pairs than many plants, fishes, invertebrates, amphibians, and protists. So what makes us more distinguished in terms of biological complexity? Is it all in our heads?

Indeed, most of the time, people point to our brains. The human brain has famously been hailed as "the most complex object in the known universe" and "the most complicated organization of matter that we know." We've alternated between various hypotheses about what makes the human brain so special over the years. But as we've shifted the goalposts, have we gotten any closer to answers?

At one point, we chalked human exceptionalism up to brain size. We long considered brain volume a proxy for mental capacities like intelligence and sentience, and essentially, the bigger the better. But the problems with this position soon surfaced. The average human brain weighs around 3 pounds. That's less than the average brains of bottlenose dolphins (3.5 pounds), and a fraction of those of animals like African elephants (10 pounds) and sperm whales (20 pounds).

Recognizing the link between larger bodies and bigger brains, we then turned to brain-to-body-mass ratios (a.k.a. encephalization quotients). On that metric we beat out cetaceans and elephants, but were defeated by a humiliating array of small mammals, including squirrels, shrews, and chipmunks.

So we turned to the computational units of the brain: neurons. But once again, we came up short: the elephant brain has around 257 billion neurons, three times that of the average human brain. Some may minimize this difference by claiming that it has to be corrected for body mass—but so-called encephalization corrections are more relevant to brain weight than to neuron counts. The brains of parrots and songbirds contain, on average, twice as many neurons as primate brains of the same mass (thus these avian brains have higher neuron densities than mammalian brains).

While the human brain contains more neurons than any other primate brain does, recent studies show that the number of neurons in human brains is not so exceptional. The human brain has been described as a linearly scaled-up primate brain, meaning that it has just as many neurons as would be expected of a generic primate brain of its size. A similar story applies to the distribution of neurons throughout the brain and other aspects of brain anatomy. For instance, relative to the rest of the brain, the sizes of the human neocortex and frontal lobes—long considered the seats of "higher" cognitive abilities—are less remarkable compared to those of other species than previously assumed.

All told, if the basis for mental powers lies in the brain, how can it be that humans, the self-designated most cognitively capable species of all, don't have much to distinguish them? Time and again, we've resorted to considering our brain as an outlier to explain our supposedly superior intellect. This makes little sense, however, in light of evolution. And yet we routinely make ourselves the exception to evolutionary rules. As one 2018 article in *Vox* exemplifies, "The human brain is so unique, and so densely complicated, it's unlikely to have developed by truly random chance."

So the search for human distinction goes on. As my undergraduate adviser once foresaw, neuroscience today has a lot of power. It's easily one of the most well-funded areas in all of science, receiving roughly $12 billion in funding from the National Institutes of Health in 2023 alone. Yet all along, characteristics that appeared to single out the human brain as an anomaly have been rethought. As cognitive scientist Gary Marcus says, "If it seems like scientists trying to find the basis of human uniqueness in the brain are looking for a neural needle in a haystack, it's because they are."

That said, we don't need to deny that the human brain might have features that differentiate us from other species, and that these unique features may alter our intelligence and consciousness. And so it is for all (brained) species. We *all* have unique brains! Octopuses are sometimes even said to have multiple brains. Two-thirds of an octopus's neurons are spread throughout her body, distributed among her eight arms. Each arm has a mini brain allowing it to act independently, giving these magnificent animals the benefit of both localized and centralized control over their actions. There might not just be *something it's like to be* an octopus, but something it's like to be nine of them.

Birds provide another useful illustration, their tiny brains long ridiculed as cognitively inferior. Yet on many metrics, some birds have proved just as cognitively capable as (if not more so than) many mammals. Their forebrains contain a region called the pallium that performs functions similar to those of the mammalian neocortex (the area thought to be involved in complex cognitive abilities and conscious experiences like pain). But whereas the mammalian neocortex is laminated (meaning cells are organized into layers), the corresponding avian pallium is nucleated (meaning cells are organized into clusters). This highlights how comparable cognitive abilities can arise from distinct anatomical structures, a point we'll pick up again later.

Our view of the human brain as an extraordinary point off the curve is likely based on the mistaken assumption that all brains follow the same blueprint. But why take the human brain and its organization as the measure of all things? As species ranging from birds and bees to elephants and octopuses have shown us, you don't need human-like brains to evolve complex minds. But do you need "brains" at all?

Neuroexceptionalism

Hiking through a forest in the Berkshires, I stop for a moment to catch my breath. It is early spring and a Carsonian silence hangs in the air. The damp forest floor is still strewn with dead foliage—mostly leaves and needles from the black birches and eastern hemlocks overhead. I position myself to retie my shoelaces, accidentally rolling a decaying log over with my boot. An unexpected burst of bright color and tiny furtive movements alerts me to the vibrant world underfoot. A familiar sight of wood lice and millipedes bore their way back into the safety of darkness. But clinging to the underside of the log is something I've never seen: a moist, banana-yellow body fanning out into dozens of elegant fractal branches. Baffled, I snap a photo and send it to several colleagues. One of them confirms: I'm in the honorable presence of a slime mold.

Slime molds defy scientific categorization. Neither animal, nor plant, nor fungus, they are actually complex, soil-dwelling amoeba (today slime molds are classified as protists, a taxonomic group biologists reserve for "everything we don't really understand," according to one who studies them). And though slime molds have no brain or nervous system, they are remarkably intelligent. Among their many claims to fame, they've gained quite the reputation as city planners.

Slime molds evolved to grow in the most efficient way possible to maximize access to nutrients, making them well-known experts at finding the fastest route to food through mazes. But a group of scientists wanted to see whether slime molds, and the yellow *Physarum polycephalum* I had a surprise encounter with in particular, could create networks as sophisticated as the transport networks humans make. The scientists, perhaps not coincidentally, were from Japan. Their rail system is among the most robust and efficient in the world, a triumph of complex engineering that took a team of talented and dedicated engineers years to realize.

In one study, the researchers placed oat flakes (a *Physarum* delicacy) in various spots on a wet surface in the pattern of the cities surrounding Tokyo. They even added areas of bright light (which slime molds tend to avoid) to simulate mountains, lakes, and other topological barriers that the trains have to steer around. They then released the *Physarum*. After just one day of adapting to the environment, the slime mold spread among the oat hubs, building nutrient-ferrying tubes into a pattern nearly identical to the actual train system of Greater Tokyo. In some ways, the slime molds' solution was even more efficient, and they figured it out in a fraction of the time it took humans.

Another study put the *Physarum* up to an even more daunting task. Imagine that a traveling salesman wants to find a route that passes all households in a neighborhood exactly once before returning to the starting point. The most accurate way to answer the so-called Traveling Salesman Problem is to measure every possible route, then determine which one is shortest. But when you get beyond a few destinations, the problem quickly gets out of hand. The exponential growth in possible solutions has proved a huge obstacle for mathematical algorithms; apparently, a few hundred points could take years to run on a supercomputer. But using the same method as the Tokyo rail experiment, researchers showed that *Physarum* took only twice as long to solve a map of eight cities as to solve a map of four cities—despite there being almost a thousand times more possible routes. In short, the slime mold easily solved the code that the most powerful computers in the world—and humans—can barely crack.

For years, there was the firm conviction that a nervous system centered in the brain was necessary for any type of cognitive complexity. As we've seen, animals like octopuses defied this logic. But more recently the conversation has shifted to include a growing diversity of lifeforms—including those with very simple nervous systems (such as sponges) or without nervous systems altogether (such as plants, fungi, and slime molds). Evidence is mounting that these other lifeforms engage with the world in highly perceptive and intelligent ways; that they, too, may possess minds as such.

Unsurprisingly, many researchers remain resolute in their conviction that these terms should be reserved for brained organisms. Such scientists have no problem admitting that organisms like plants *behave* cleverly, but they consider projecting "human" mental capacities onto them a bridge too far. For instance, in an interview with *Smithsonian* magazine, Oxford paleobiologist Richard Fortey critiques the rhetoric surrounding trees as "so anthropomorphized that it's not really helpful." He explains, "Trees do not have will or intention. They solve problems, but it's all under hormonal control, and it all evolved through natural selection." In other words, plants exhibit no agency or learning; their behavior is instead largely driven by instinct. Interpretations beyond that, he cautions, risk leading people to "immediately leap to faulty conclusions, namely that trees are sentient beings like us."

At least part of the skepticism surrounding plant intelligence traces back to an influential book, *The Secret Life of Plants*. Published in the 1970s, its authors claimed that plants could sense human emotions, based on experiments in which they hooked up plants to polygraph machines. The scientific community derided the book as pseudoscience, which stymied further research into plant cognition for decades. But a recent resurgence in research by scientists using more validated methods is reigniting the conversation.

One of those scientists is the evolutionary ecologist Monica Gagliano, a friend and collaborator. In 2014, she conducted a series of studies to test whether plants, like animals, could demonstrate a basic type of learning called "habituation." Habituation enables organisms to attend to meaningful stimuli in their environments while ignoring those that have proved irrelevant and innocuous.

Monica works with a species called *Mimosa pudica*. Gardeners call *Mimosa* "the sensitive plant" because if you touch them lightly or otherwise disturb them, they quickly fold their tiny leaves into what looks like a frightened or defensive recoil.

In one experiment, Monica custom-built an apparatus that dropped these sensitive plants from a short height. Initially, the *Mimosa* retracted and curled their leaves, a typical threat response. But after repeats, the plants stopped reacting. In other words, they habituated to the drop. They learned that it wasn't a threat.

Maybe, as skeptics suggested, the plants merely stopped reacting because they were fatigued. Anticipating this criticism, Monica introduced a new threat: shaking the plants. When the pots were shaken, the leaves closed up again—something they can't do if depleted. Perhaps even more astonishingly, the plants also remembered. When the familiar vertical drop test was repeated a month later, their leaves remained unruffled.

Learning and memory are the best descriptors for these results, but this language has caused quite a stir in the scientific community. Because plants are brainless creatures, opponents say, they can't learn and remember. Indeed, plants (together with fungi and several animal taxa like corals and sponges) lack neurons altogether. Most are modular, characterized by an iterative, indeterminate mode of growth. But this is one of their strengths. With no central, irreplaceable organs, most plants can survive losing 90 percent of themselves, and many species can grow and survive independently from broken pieces or cuttings.

The prevailing view that plants aren't intelligent because they lack a nervous system is sometimes called "neuroexceptionalism." The ecophilosopher Derrick Jensen likens it to a tautology. In math, a tautology is "a logical statement in which the conclusion is equivalent to the premise." It's similar to circular reasoning. If we define something like intelligence as a phenomenon that requires a brain, then by definition those species without brains and neurons will lack this quality. It's a logical problem, because it doesn't make sense to define something in terms of itself! Many debates over the intelligence of animals or plants or even artificial intelligence "are thus revealed as endless—and ultimately, rather pointless," according to the writer James Bridle, "because they are debates about the meaning of words rather than the being of things."

Indeed, Monica claims her opponents don't have problems with the data but with the words she used. They ask, is this really learning and memory? Or is this just metaphorical shorthand, bridging the gap between the familiar world of animals and the unfamiliar world of plants? Monica insists that she is not using metaphor. That "learning" and "memory" are the best words—both processes that enable past experiences to be brought to bear on the present—and to use different words would be the error.

Scientists rightly strive to operationalize terms, but should those terms be initially defined in a way that precludes cross-species comparisons (such as requiring a brain for learning or memory)? Don't these findings give us an important opportunity to rethink what learning and memory really are and how they work—to study how they can arise by other means, in other structures, and in other species? This less anthropocentric science would tell us about minds generally, rather than how animal or plant minds relate to human minds. It would stop placing the human mind at the center of the cosmos with all other minds orbiting around it.

Function over Form

Plants' cognition should be considered in their umwelten, which yield distinctive ways of interacting with the world. Plants face many of the same problems as animals (finding energy, reproducing, evading predators), though they differ significantly in their approach. For a start, plants are rooted, sessile organisms. While they can stretch toward sunlight and bend with gravity, they can't readily flee or escape. They've thus evolved especially sophisticated ways of defending themselves and warning others.

Many species of plants, scientists have learned, recognize specific predators and mount complex, targeted defenses in response. For example, when the flowering mouse-ear cress senses the vibrations caused by caterpillars munching on them, they release oils and chemicals to repel these hungry herbivores. Some plants respond to being eaten by insects by producing a blend of compounds that attract carnivorous enemies of those same insects. When they're attacked by mites, for instance, lima bean plants emit a chemical SOS that recruits larger predatory mites—one specific to this particular kind of threat.

Plants also exhibit richly complex communication networks with other plants. They send airborne chemical messages that warn one another of impending danger, or do so through the fungal "wood wide web" connecting their roots. These messages provide specific information to neighboring plants, who prepare for the perceived threat accordingly. Remarkably, plants can also respond to messages emitted from other plant species.

While they may have a limited capacity for movement, plants have astounding abilities to perceive and react to the world around them. They have senses analogous to our five, and then some. In addition to hearing, touch, and taste, they can sense electromagnetic fields, gravity, and the presence and properties of water. We've all seen houseplants salute the sun each morning, because plants sense light and grow toward it. But their sense of touch is especially impressive: research has shown that plant roots actually shift their direction of growth *before* hitting a rock or another obstacle. We now know that a single root apex can detect at least twenty different chemical and physical parameters, many of which are invisible to us.

Plants appear to recognize self from nonself, and can even discriminate relatives from unrelated plants. Experiments reveal that searocket plants (a succulent found on North American beaches) can distinguish between kin and unrelated searockets. When planted with strangers, the searockets greatly expand their underground root system; with relatives, they leave more room for kin roots to get nutrients and water, thereby presenting less competition. Evidence for plants' altruistic tendencies initially seemed outlandish, but many studies since have lent further support. In addition to warning one another of danger, trees share nutrients via their root networks at critical times. It has been suggested that these networks facilitate recovery and succession of forests after disturbance.

Some scientists have recently argued that plants can experience something like pain. When anesthetized, Venus flytraps don't respond as they normally do to insects crossing their threshold. Similarly, dosed-up *Mimosa* plants lose their leaf-closing response to perturbations, only to return to normal after several hours. When injured or stressed, plants also produce their own compounds that are anesthetic to us, such as menthol and ethylene. "Plants are not just robotic, stimulus-response devices," explains plant cell biologist František Baluška in an interview with *The New York Times*. "They're living organisms which have their own problems, maybe something like with humans feeling pain or joy. In order to navigate this complex life, they must have some compass."

No longer confined to the fringes of new age belief, this growing understanding of plants as active beings with rich, sensual lives is turning fundamental assumptions upside down about what separates plants from animals. Yet why is it that so many of us still balk at such conclusions? Why do they seem bizarre and unscientific? Because our understanding of plant intelligence is limited, we tend to dismiss it, or study it in ways that rule it out. On the other hand, according to Indigenous botanist Robin Wall Kimmerer: "If we would embrace the possibility of plant intelligence and investigate it without any anthropocentric bias, we might be surprised by what we learn."

~~~~««llltll»»~~~~

Aristotle's *scala naturae* ranked plants below animals on the basis of their apparent immobility. But anybody who has seen a time-lapse nature documentary probably knows otherwise. Plants move, just on a slower timescale than animals like us.

Charles Darwin was one of the first scientists to use time-lapse methods to investigate the vitality of plants. In a book titled *The Power of Movement in Plants*, published with his son Francis in 1880, he recounted their joint experiments investigating the sensory capacities of plant roots. Together, they introduced the controversial notion that the tip of the root acts like the brain, receiving sensory input and directing movement—what they called the root-brain hypothesis. It's worth noting that plant cells share features with all biological cells, including neurons. To name just a few: plant cells produce spikes of electrical potential, their membranes harbor ion channels, and they contain neurotransmitters such as serotonin, dopamine, and glutamate.

But more importantly, what if we stopped measuring capability by morphology (the branch of biology that deals with the *form* of living organisms) and instead looked at *function*? Remember that evolution makes clear how completely different structures can arise to serve a similar purpose (most fish have gills while birds and mammals have lungs, achieving a similar function—exchange of gases with the environment—by different means).

Because advances in our understanding of cognition conventionally come from humans and other animals, it's implicitly accepted that intelligence and consciousness (and anything related to "the mind," really) are based on neuronal processes. If the form (nervous system) is missing, we assume the function (thinking) must also be absent. But this is just that: an assumption, one frequently masquerading as an obvious truth.

Thinking is what allows one to use information flexibly in changing environments and to make sound decisions honed through experience. To ask "Is that thinking?" is ambiguous because it all depends on how we define it and, namely, whether we do so in human-centric terms. A clearer question might be something like "How did evolution solve this particular function with different morphology?"

Might evolution have solved the function of thinking in a different way for plants than animals? Plant scientists would be the first to admit that the mechanisms of intelligence in plants and animals differ. Such is the case for many processes for which we have similar words (e.g., respiration, circulation, digestion, immunity, reproduction, growth, development, and others). Yet when it comes to *minds*, we tend to invoke a hierarchy—the Great Chain of Being, which persists one way or another to this day.

What It's Like

Now the big question: Are these other forms of life conscious?

Plant biotechnologist Devang Mehta says, "The answer, unreservedly, is 'no.'" He is one of many scientists who emphatically argue that plants (let alone fungi, slime molds, and other nonneural organisms) cannot be conscious because consciousness requires a structurally complex nervous system. Lincoln Taiz, a well-known plant physiologist, similarly has little patience for the notion that plants can experience conscious states like suffering: "No brain, no pain," he roundly states. When all is said and done, this view holds, there's *nothing that it's like to be* a plant.

Debate still rages around using the word "consciousness" in the nonhuman context, let alone the non*animal* context. Yet to my knowledge, few plant cognition scientists argue that plants are definitively conscious. Instead, they advocate a more humble posture—to not foreclose the possibility that plants have minds just because they lack neurons. As scientists, they ask *what if?* Perhaps we understand consciousness so little because we've been going about it all too narrowly; perhaps opening our minds to other species' minds will help us understand minds (including our own) more fully.

I don't think it's any exaggeration to say this was a life-changing realization. Given that we are steeped in a long tradition of human exceptionalism, it's easy to assume that our experience of life is what defines consciousness—that our brain's processes are the pinnacle of mind. I was drawn to our primate cousins because they challenged this supposed human-animal divide. I then realized that minds need not resemble ours to be comparably complex, only to discover that limiting these arguments to the animal kingdom is another form of anthropocentrism.

Such is the gist of many conversations I have had over the past two years as part of Harvard's Plant Consciousness Reading Group, led by my friends Natalia Schwien and Rachael Petersen. It has been especially intriguing to note the parallels between contemporary debates on plant minds and historical discussions questioning the intelligence and consciousness of other animals. Expanding on this, I find myself wondering why we keep redrawing the limiting line for mindedness across lower and lower rungs of the *scala naturae*. At what point do we consider abandoning these "lines" altogether?

Scholars in the rapidly growing field of basal cognition, like Pamela Lyon, argue that evolution already laid a solid foundation for capacities typically considered "cognitive" 500 to 650 million years ago, well before nervous systems appeared. "Perception, memory, valence, learning, decision-making, anticipation, communication—all once thought the preserve of humankind—are found in a wide variety of living things," writes Lyon, "including bacteria, unicellular eukaryotes, plants, fungi, non-neuronal animals, and animals with simple nervous systems and brains." When does it become more parsimonious to assume that awareness and sentience are fundamental and ubiquitous aspects of the natural world? Though still a minority view, the belief that there is some form of consciousness in everything is gaining more credibility among Western philosophers and scientists (many believe this view—generally known as panpsychism—offers the best hope of cracking the hard problem of consciousness). But as we'll explore later, countless Indigenous, animist, and pagan traditions have long regarded the sentience of the more-than-human world as self-evident—an awareness rooted in the experience and cultivation of reciprocal relationships.

Theories about consciousness are complicated, and I wouldn't insist that the possibilities I've raised here are all true. I would insist only that human exceptionalism masks our ability to judge these possibilities without bias, and that they are at least as plausible as the weirdly inconsistent notions that currently pass for common sense. Of course, it's a stretch of the imagination to think about what consciousness might even mean for these other organisms. But that's surely no foundation for refuting its existence. Why should we doubt, downplay, or deny something in other species that we barely understand in ourselves? That I barely understand in *my*self? How might reversing the burden of proof—assuming that sentience is ubiquitous in all its exquisitely diverse forms—transform our science? And our relationship to the natural world more broadly? Learning to ask these questions expanded my own way of being, filling my every day with a renewed sense of childlike wonder, curiosity, and openness to connections with all kinds of life. Living in the world in this expanded way is a permanent gift.

We'll end this chapter where we began, with Nagel's question: What is it like to be a bat (or a chimpanzee, trout, or pea plant for that matter)? Was he right that we will never be able to answer such questions—that they lie beyond our ability to conceive? Then again, by the same reasoning, we can't really know what's going on in each other's minds either. But this should not stop us from trying (indeed, it does not stop us from trying to relate to other human beings every day). Some maintain we may never solve the so-called problem of other minds. It is, however, a problem that deep relationships with others can help us imagine solutions to.

CHAPTER SEVEN

Relationship Matters

THE MOST INTIMATE MOMENT I EVER SHARED WITH A WILD animal caught me by surprise. It was Thanksgiving, and I was spending it with a troop of desert chacma baboons in Namibia. The Namib Desert sun in November is scorching hot, with temperatures easily exceeding 100 degrees Fahrenheit. Trudging through a dry riverbed carrying several liters of water on your back quickly gets tiring. Luckily, the baboons know this arid climate well, so they make their way to a rocky shelter at peak sun, and we all get some relief. The baboons situate themselves in various recesses and cavities of the grotto. My colleague Elise Huchard and I install ourselves on opposite sides, repurposing our backpacks as headrests as we sip some water. A hot breeze whispers through the grotto, carrying the distinctive aroma of nearby African myrrh shrubs.

After a while, it falls eerily silent—desert silence, the kind of silence that makes you realize just how noisy everywhere else in the world is. Only then does it occur to me that everybody has fallen asleep. All two dozen or so baboons, and also Elise. I am not a napper (it may be the skill I envy most in others), so the all-too-familiar feeling of being the only one awake sets in.

Then someone stirs to my right, signaling I'm not alone. A juvenile female in the alcove beside me rolls over nonchalantly. She lets out an audible sigh as her eyes search the grotto. Soon enough her gaze reaches mine. We blink at each other in the shadowy light, knowing it's just us two left behind in the waking world.

Then it hits me: this is the first time I have ever been truly alone with a wild primate. I had been with baboons one-on-one before, but the presence of other group members (or a human colleague) was always palpable, even if they weren't close by or visible. It strikes me how very different this now feels.

My sleepless baboon companion fiddles with some pebbles in the alcove, poking and prodding them with her fingers. Their clattering makes a strangely satisfying sound. She pushes a few off the alcove's ledge, glancing sideways at me for my reaction. I attend to her not as a scientist trained to decode behavior but as a fellow animal engaged in the immediacy of this shared moment. Periodically, we both turn on our backs and look up, eyes scanning the open sky. Time stands still for a little while.

It's hard to put into words why this experience was so profound. On the surface, I've had much more "extraordinary" encounters with baboons—grooming sessions, tugs-of-war, and reconciliatory interactions. By comparison, nothing obviously remarkable happened here. And yet this is the experience I recall most vividly, the one I return to most often. I had studied primate cognition and emotion for years. But never had I been so convinced of the minded life of a baboon—one as full, nuanced, and idiosyncratic as my own—as I was in that suspended moment of reciprocal encounter. In truth, it wasn't clear to me exactly what she was thinking or feeling. But it was clear to me that she *was* thinking and feeling. That there was "someone home," as my friend the primatologist Barbara Smuts likes to say. Nothing seemed more obvious to me than that.

Encounters

In a landmark paper titled "Encounters with Animal Minds," Barbara Smuts recounts her time living with and studying wild savanna baboons near Kenya's Lake Naivasha. For years, she traveled among them, from dawn until dusk, often with no or minimal contact with other humans.

In the beginning of her research, she had to convince the baboons that she was not a threat. Initially, this meant approaching the troop from a distance and halting whenever they began to flee, biding her time until they relaxed before inching nearer again. She found that progress was slow, until she started noticing more subtle baboon signals—like mothers calling and directing stern looks to their infants when she got closer. Smuts learned to stop approaching *before* the baboons got nervous. And soon enough, she was allowed to get very close, eventually moving freely in their midst.

The accepted scientific term for this process is "habituation," implying that the animals get used to the researcher while the researcher (essentially) stays the same. But in reality, Smuts clarifies, nothing could be further from the truth. To gain the baboons' trust, she had to change almost everything about herself—how she held her body, moved, directed her gaze, used her voice, attended to her surroundings, and more. "I was learning a whole new way of being in the world—the way of the baboon," she writes.

Remaining a detached, "neutral" observer is often considered fundamental to good science. Accordingly, most fieldwork training manuals instruct you to minimize any influence you might have on animals' behavior. Likewise, Smuts was taught that if a baboon approaches or tries to interact, she should ignore or slowly move away from them. But as she quickly learned, ignoring the solicitations or even proximity of baboons is rarely a neutral act, because they regarded her as a sentient subject—not a neutral object—in their environment:

"Although ignoring the approach of a baboon may at first sound like a good strategy, those who advised me to do so did not take into account the baboons' insistence on regarding me as a social being. After a little while, I stopped reflexively ignoring baboons who approached me and instead varied my response depending on the baboon and the circumstances. Usually, I made brief eye contact or grunted. When I behaved in this baboon-appropriate fashion, the animals generally paid *less attention* to me than they did if I ignored them" (emphasis added).

That is, contrary to convention, Smuts had to lean into (not avoid) a responsive relationship if she really wanted to understand baboons and let them go about their daily business. This kind of mutual engagement with the animals offered insights that were impossible to obtain in any other way:

"Being treated like a fellow baboon proved immensely useful to my research, because I experienced directly critical aspects of baboon society. For example, I soon learned that the baboons' most basic social conventions entail acknowledgement of relative status through respect for personal space. In general, each baboon has a small invisible circle around him or her that a lower-ranking animal will rarely invade without first signalling intent (usually by grunting) and receiving from the other an indication that it is safe to approach (usually a reciprocal grunt and/or the 'come hither' face)."

Smuts soon became aware of even subtler social dynamics, in that each troop member's personal space can shrink or grow depending on the circumstances. When a male courts a female, for instance, her personal space tends to expand (a shift to which the male must be sensitive if he wishes to woo her successfully). And when a dominant baboon attacks a subordinate baboon, the subordinate's personal space in relation to the dominant expands until the two have reconciled or tensions have otherwise subsided. Again, Smuts's formative insights into baboon sociality were not achieved by remaining a cold, disinterested observer; rather, they relied on relating to the baboons in a way that recognized their subjectivity and individuality.

Smuts is part of a tradition of women primatologists—including Jane Goodall, Dian Fossey, and Birutė Galdikas—who made foundational contributions to understanding our primate relatives because they encountered these animals first and foremost as social beings rather than as objects of scientific inquiry. Their approach is distilled in the Japanese word *kyokan*—meaning to empathize or feel one with another—a method also adopted by Japanese primatologists that yielded profound insights into primate societies. Although forming relationships based on reciprocal accommodation and trust is sometimes taken to be at odds with being a careful scientist, this work gives us reason to believe that such relationships are the basis of genuine understanding. A phrase from C. S. Lewis comes to mind: "You can't study men; you can only get to know them."

As Individuals

Ask any primatologist about their favorite individual animal they work with. They will almost always have an answer at the ready. There's an ongoing joke in my field that it's typically not the researchers who select. Instead, the "chosen one" is the researcher. You may recall that Winter, the brown capuchin monkey who lip-smacked at me on my first day of work, initially approached *me* to extend this welcoming gesture. And naturally, she became my favorite capuchin after.

Living over prolonged periods with animals like baboons, you come to know each one as a highly distinctive individual. At first, you may identify them via physical traits—body sizes; ear, tail, or muzzle shapes; and distinctive scars or wisps of fur. But over time, you come to perceive individuals as a constellation of habits and preferences—each with a specific gait, characteristic voice, style of social interaction, favorite food, ideal resting spot, and other distinguishing traits. Further, each baboon knows you as an individual and approaches their relationship with you slightly differently, a sure way to tell them apart. This mosaic of mannerisms—baboons' whole bearing and character—is too idiosyncratic *not* to see, so much so that it can later be difficult (for instance, when training other researchers) to recount the "obvious" physical characteristics that differentiate them. You know who's who from a mile away, but can't explain why.

Animal personality—individual differences in behavior that are stable over time and context—has been described in organisms as diverse as sea anemones, mice, elk, spiders, cockroaches, raccoons, hermit crabs, and lizards, suggesting it exists across the entire animal kingdom. Yet research in this area was once seen as a "soft science" that risked committing the scientific sin of anthropomorphism, which is one reason why the field got off to a slow start (human psychologists had studied personality for at least a century prior). Individual differences were long minimized and considered troublesome because they got in the way of "averages" or adaptive optimums of behavior. But anybody with two or more companion animals can easily distinguish differences in their characters, and scientific understanding eventually caught up. Recognizing such variety is central to moving beyond the anthropocentric tendency to consider other animals as homogenous units, who express the inborn or acquired blueprint of a species. And this is to say nothing of the field's fundamental evolutionary implications (after all, individual variation is the raw material on which natural selection acts).

My interest in animal behavior naturally made me curious about whether personality played a role in chimpanzees' conflict-resolution habits. In one longitudinal study, my colleagues and I found that some chimpanzees consistently reconcile more than others, and do so more quickly following their disputes. These differences were not attributable to factors like an individual's rank or sex; instead, the propensity to resolve conflicts reflected an individual's personality (as well as their particular relationship with the opponent). A few years later, we looked at whether reassuring others in distress (consolation) also differed reliably across individuals. Again, we found striking personality differences in the tendency to console. Given the link between consolation and empathy, our findings suggest that some chimpanzees are more empathic than others, perhaps something that shouldn't surprise us given well-known individual differences in human empathy.

For a seemingly endless variety of traits—whether empathic or aloof, audacious or cautious, aggressive or meek, friendly or antisocial—personality in chimpanzees is the rule, not the exception. Most recently, my student Zoë Goldsborough found that chimpanzees even express stress in highly idiosyncratic ways—what's "normal" for one chimpanzee might be an indication of poor welfare for another.

When you consider species beyond our close primate cousins and cherished companion animals, the variety of personality types increases exponentially. As Barbara Smuts reflects, "My awareness of the individuality of all beings, and of the capacity of at least some beings to respond to the individuality in me, transforms the world into a universe replete with opportunities to develop personal relationships of all kinds."

In 1938, a British naturalist and musician named Len Howard purchased a plot of land in Sussex and began constructing a home. She later called it Bird Cottage because she shared it with robins, sparrows, blackbirds, great tits, thrushes, finches, and other birds living in the area. Howard believed the best way to study birds was in a naturalistic setting instead of in laboratories (as was standard at the time). So she kept her windows open and let the birds come and go as they pleased.

Howard lived with the birds for many years. She got to know them intimately as individuals, for they quickly learned they could trust her. Unlike traditional studies of bird behavior in which birds are often frightened, Howard believed that gaining the birds' confidence was crucial to understanding their intelligence: "I have no doubt the Blue Tits and Robins would not have behaved intelligently if they had feared my presence. Often bird behaviour is judged when the bird is panicked with fear of the watcher. But many humans would prefer not to have an intelligence test when they or their young are in probable danger of immediate death."

Howard's book *Birds as Individuals* is filled with evidence (including delightful photographs) of birds at ease in her presence: perched on her shoulder, flopping into her lap, and hiding from rivals in her hair. Close contact and attention taught Howard to read the birds' bodily, facial, and vocal expressions (and they, hers). She soon realized birds express themselves intricately through their beaks. She came to learn that some birds are habitually bold and outgoing, while others remain shy and fearful; some are well-mannered and gentle, whereas others are demanding and testy. Birds practice distinctive territoriality displays, play styles, and mating rituals, and they have specialized quirks regarding food and roosting. Some are highly creative, singing songs with great innovation, while others are less original and instead prefer to mimic conspecifics.

Howard's research took place at a time when people underestimated avian intelligence (as she once exclaimed, "Humans who are over-conceited about the cleverness of mankind should live with

Great Tits so as to see things in proper proportion!"). Her work challenged the dominant perspective that birds mindlessly exhibit a set of automatic, preprogrammed behavioral responses.

One of the most fascinating examples of bird cognition in *Birds as Individuals* features a Great Tit named Jane. Jane became visibly alarmed when her fledglings first flew near the windows of Bird Cottage, knowing that young, inexperienced birds often collide with glass. Recognizing the danger, Jane grabbed some food and recruited her fledglings to an adjacent perch, placing herself on the opposite side of the glass. Jane then moved momentarily outside the window to show the food but, before they could take it, returned behind the windowpane so that they would peck the glass again. Jane repeated this clever demonstration several times until her fledglings clearly understood the lesson, as seen in their behavior in the days that followed. As Howard notes: "There was no trouble when they afterwards followed their parents into the room because Jane had taught them to know glass; they were cautious when approaching the windows, found their way out of the open ones and never banged into those that were closed. What else except reasoning power could have accounted for Jane's action?"

This example is especially remarkable because teaching is among the qualities typically considered human-unique. No doubt some will be skeptical of such claims, giving them the pejorative label of "anthropomorphic" or "anecdotal." And yet, who are we to second-guess Howard's expertise when we have not observed and engaged with birds as thoroughly as she did? Might her extensive experience studying and living with birds put her in a position to know things about them that the rest of us do not know, to perhaps *literally see* things about their lives and personalities that we cannot see?

Reading this book also opened *my* eyes to bird behaviors I had never noticed before. For instance, I recently observed a robin fledgling struggling to take flight while his siblings swooped around effortlessly. Sensing his fear, the mother robin grabbed a worm and flew close to him, then quickly swerved to a perch about a meter away. She repeated this tempting gesture several times, and I watched as her hesitant son inched closer each time, torn between fear and the urge to fly. Finally, he took the leap, flying to the perch and earning his reward. The mother then found another worm and repeated the process, gradually increasing the distance to encourage his progress. It's incredible how our expectations guide our observations (learning about avian teaching led me to see it firsthand), and how attending to the nuances of individual behavior and personality uncovers a world of complexity. As Darwin once wrote, "It is a significant fact, that the more the habits of any particular animal are studied by a naturalist, the more he attributes to reason and the less to unlearnt instincts."

The divide between human and nonhuman beings, so central to the dominant culture of human exceptionalism, diminishes as individual identities come to light. Len Howard gained remarkable insights about birds' mental lives—a privileged understanding that came to be through intimate relationships that allowed her to get to know them as individuals. So why is this way of relating to other species today often deemed controversial, or even taboo, in a scientific context?

The Empathy Taboo

The whole notion of the detached and uninvolved scientific observer has a long history—one deeply intertwined with the history of human exceptionalism. At its source is the modern assumption that removing the "self" from research frees knowledge from the taint of subjectivity and renders it purely objective. As the British philosopher Mary Midgley points out, however, maintaining distance is not the same as calling for objectivity. As she explains in a 2001 *Nature* paper, "objective" means unbiased and impartial. Detachment does not necessarily achieve this goal. Instead, reacting to others without emotion introduces a different kind of bias, as it treats others "as a lifeless object, not as a subject." In other words, it confounds "objectivity" with "objectification."

The belief that we should treat everything we study as a mindless, inanimate object—what my colleagues and I call "the empathy taboo"—has long survived the Cartesian and behaviorist ideologies out of which it has grown. But people who break the taboo by responding to other species as minded, relational individuals do not compromise their research or make it any less insightful or rigorous. On the contrary, their empathy allows the animals to express themselves freely, affording a unique and invaluable window into their lives.

"Someone who sees children merely as unconscious physical objects is not going to be able to observe them very intelligently," writes Midgley. She acknowledges that there is no utterly neutral vantage point from which we can experience and describe the world, making true objectivity impossible in practice. But she insists that "we have to deal with this difficulty by watching for the obvious sources of bias, not by suppressing our natural responses altogether." That's why scientists studying human children typically spend time building rapport and trust with the subject before the experiment. Likewise, ethnographic fieldwork relies on researchers developing empathic connections with the human participants and cultures under study. Why shouldn't the same be true when researching other living beings? It's another self-fulfilling prophecy; by relating to other species as mere things, we limit our understanding of them. As the philosopher Stanley Cavell puts it, "Here, what we do not know comprises not our ignorance but our alienation."

Jeffrey Mogil is a neuroscientist studying pain perception at McGill University. One day he and his colleagues stumbled upon a surprising result: rats and mice behaved differently in experiments depending on whether a man or woman conducted the study. Up until then, the phenomenon had remained somewhat taboo: "Scientists whisper to each other at conferences that their rodent research subjects appear to be aware of their presence, and that this might affect the results of experiments," Mogil disclosed in an interview, "but this has never been directly demonstrated until now." When Mogil's team reanalyzed their past work, they found that mice routinely showed a higher threshold for pain when handled by men than by women (it turns out that laboratory rodents tend to be more stressed when they can smell men, and this stress-induced reaction makes them less sensitive to pain).

Biomedical studies strive to keep relevant environmental variables under strict control, yet (in part because of the empathy taboo) this typically doesn't include aspects of the human-animal relationship, like the experimenter's gender. And as Mogil and colleagues point out, ignoring this variable may be a primary reason for the lack of replication in animal research, yielding major implications for all kinds of physiological and behavioral studies. Indeed, one 2022 study found that ketamine appears to be more effective as an antidepressant in mouse experiments when they are conducted by a man.

The anthropologist Donna Haraway once wrote, "Beings do not pre-exist their relatings." And as these examples show, the relationship between scientist and subject is a relevant variable. Yet the human dimension of these studies is traditionally overlooked, except as the target of "Clever Hans" objections that human interference of any kind must be avoided.

Clever Hans

Most students of animal behavior have heard the cautionary tale of Clever Hans, a performing horse in Berlin who drew worldwide attention with his apparent ability to solve complex mathematical problems. When his teacher and owner, Wilhelm von Osten, read out an equation, Hans would tap the correct answer with his hoof. Hans reputedly not only grasped addition, subtraction, multiplication, and division but could also recognize musical notes and intervals, tell time, and keep track of the yearly calendar. "If the eighth day of a month comes on Tuesday, what is the date for the following Friday?" von Osten would ask. And Hans would correctly tap eleven times.

Public interest in the spectacle led the German board of education to set up a commission to determine whether the claims made about Hans were valid. After over a year of close scrutiny, the commission concluded that there was no hoax involved. And the show went on!

It would be several more years until the careful experiments of a psychologist named Oskar Pfungst solved the mystery. Pfungst found that Hans tapped the correct answer only when he could see his questioner and when the questioner knew the answer. He concluded that Hans was responding to the unconscious signals of his trainer, whose posture and facial expressions subtly shifted when Hans approached the final, correct tap.

The Clever Hans effect—when an experimenter's involuntary cues affect a subject's performance—is generally taught as a precaution against so-called experimenter effects. While it's essential to implement careful controls to ensure you're measuring what you think you are, people often try to minimize Clever Hans effects by distancing themselves from the beings they study—if not by studying animals in an isolated apparatus, then by wearing a face shield or turning one's back while studying them. In other words, by minimizing the potential for relational engagement. Yet as we've seen, this can create problems of its own, because it risks objectifying other animals who might find these theatrics (and being ignored) all rather strange.

According to Pfungst's "killjoy" account of Hans's behavior, horses (among other species) succeed in seemingly complex cognitive tasks using simple behavioral rules. Such deflationary accounts tend to portray other animals mechanistically; using Pfungst's own words, "instead of the much desired intellectual feats on the part of the horse, there was merely a motor reaction to a purely sensory stimulus." By showing that Hans's performance was not the mark of human symbolic thinking, Pfungst deemed this horse's behavior a facade for a much simpler feat, one that is decidedly *not* clever: "The large number and the irregularities of the errors showed that there was *no manner of intelligence* involved" (emphasis added).

When I first learned about Clever Hans, it struck me that Hans was able to do something quite remarkable. He learned to read the inadvertent cues of his German trainer, cues that remained invisible to most human audiences (including von Osten himself, not to mention a committee of discerning experts). In this light, Hans's active engagements with von Osten may be seen as evidence of highly attuned and attentive interactional skills (with a member of another species, no less!). How ironic that one of Hans's greatest competencies gets framed as "killjoy" evidence that he wasn't so intelligent after all! And that one of the most interesting, illuminating dimensions of the research—the embodied understanding that emerged between species—became emblematic of what is to be avoided when trying to understand animal minds.

Embodied Minds

In Western philosophy, phenomenology—the philosophical study of experience—most powerfully called into question the notion of an impersonal, objective approach to understanding the world. Maurice Merleau-Ponty, a French phenomenologist, famously argued that our everyday perception is not detached and objective but is inherently participatory and embodied. One cannot observe others impartially "from the outside" because, from the standpoint of our lived experience, the mind and body are inseparable.

To illustrate this point, consider our relationships with companion animals. Amber was my family's second Shetland sheepdog, and we grew up alongside each other until I left home for college. Through our close bond, I came to understand her emotions, intentions, desires, and personality (and she mine). I achieved this understanding not by accessing some otherwise hidden inner "mind," but by being immersed enough in her embodied life to become deeply attuned to her way of being in the world. As the sociologists Arnold Arluke and Clinton Sanders put it: "The understandings we derive in our encounters with companion animals are found largely in our connections to them built up over the course of the routine, practical, and empathetic interactions that make up our shared biographies. In other words, through understanding the bodies and behaviors of companion animals we actively construct a view of their minds."

In its emphasis on shared bodily experience, phenomenology rejects Cartesian and behaviorist notions of the hidden and private nature of others' mental states. For Merleau-Ponty, the so-called problem of other minds is bizarre from the standpoint of our everyday lived experience: "A face, a signature, a form of behaviour cease to be mere 'visual data' whose psychological meaning is to be sought in our inner experience." Instead, "the mental life of others becomes an immediate object, a whole charged with immanent meaning." The *problem* arises only if we cut ourselves out of the picture—if we assume that understanding another mind is possible without acknowledging our role as participants in the situation. For phenomenologists, living bodies are semiotic and replete with meaning, often carrying as much significance as words (under a phenomenological perspective, language itself is material and bodily). So the perception of other minds is not based on inference—deducing "inner" thoughts from "outer" behavior—but on something far more immediate. To perceive a body is to perceive a mind.

If this all seems counterintuitive, it's likely because of a persistent dualist perspective—one that differentiates the thoughtless, physical body from the thoughtful, immaterial mind. Most scientists and philosophers today openly reject Cartesian mind-body dualism, but it continues to shape many of our intuitions about the way the world works. And as we've seen, this dualistic view often has the practical consequence of portraying entities in Nature as mechanisms, objects, and resources to be exploited rather than revered. After all, the *scala naturae* and Great Chain of Being are also ladders of de-embodiment (where those at the top are pure mind and spirit, while those at the bottom are pure body and matter). As the writer Melanie Challenger argues, the notion that our bodies don't matter is deeply rooted in human exceptionalism and a profound struggle with our own "animality."

Strange Kinship

For phenomenologists like Merleau-Ponty, reality is grounded in the basic idea that others, too, perceive and experience the world as embodied minds or subjects, resulting in a multitude of subjective perspectives. Our understanding of the world is not solely individual. Instead, it is intricately entwined with the experiences of others, whether humans, dogs, or any number of other sentient subjects, forming a shared sense of being. For instance, over time my dog Amber and I came to anticipate our daily routines, like mealtimes and walks, demonstrating a set of mutual expectations. Her acute senses and instincts shaped my perception of the environment, alerting me to things I otherwise wouldn't have noticed—approaching strangers, inconspicuous mushrooms, evidence that deer had recently congregated in our garden—and creating a shared sense of safety and awareness. Instead of striving to divorce knowledge from the influence of subjectivity in pursuit of pure objectivity, phenomenologists maintain that these intersubjective experiences (i.e., experiences arising through interactions between subjects) are integral to our understanding of reality.

That sense of mutuality is embedded in the word "consciousness" itself, which originally derived from the Latin *conscius* (*con* meaning "together," and *scire* meaning "to know")—"knowing with" or "having joint or common knowledge with another." The emphasis on understanding reality in terms of relationships between entities (rather than entities in themselves) also finds an intriguing counterpart in modern physics. According to quantum theory, a subatomic particle can be present in multiple locations simultaneously, existing as pure potential until it is measured—meaning observed by a conscious mind. Only then does it manifest in our familiar reality, with defined coordinates in time and space. The mere observation of entities changes them, suggesting that entities can exist only in relation to other things. Because matter's existence may rely on a perceiving subject, quantum mechanics challenges the idea that science can discover a reality that exists independently of our observation of it. Under this relational quantum view, the very notion of a neutral observer loses its meaning.

I often wonder whether the prohibition on intersubjectivity and relationality in science is the reason why endeavors to *demonstrate* specific capacities in other species often fall short. Take the capacity for language, for example. Studies usually involve training other animals on *our* languages or leveraging new technologies (e.g., machine learning and robotics) to listen to and translate *theirs*. A more fruitful approach grounded in intersubjectivity might involve cocreating a new *shared* language—with researchers as active, engaged participants—much like how many people living with companion animals naturally do over time.

Merleau-Ponty once described the bond that results from these intersubjective, intercorporeal engagements with other species as "strange kinship." Strange kinship is based on "shared embodiment in a shared world, even if the style of body and the style of inhabiting that world are radically different," explains the philosopher Kelly Oliver. In this way, strange kinship acknowledges differences between humans and other animals but does not allow these distinctions to create privileges. Rather, differences in perception and modes of being are celebrated as various capacities of shared, sensory, embodied experience. The relationship between humans and other animals is not hierarchical but lateral, rooted as it is in our shared bodily natures.

Ample evidence shows that when humans have extended opportunities to live alongside other species, profound intersubjective relationships can develop—even among those with whom we might not traditionally think lasting bonds are possible. Daily I encounter viral Instagram, TikTok, and YouTube footage documenting these surprising friendships. A human diver forms a trusting relationship with wild sharks, enabling her to remove hooks from their mouths. A gardener nurtures a wingless honey bee, resulting in an extraordinary companionship. The more unexpected the interaction—the "stranger" the kinship—the better, so it seems. The global popularity of the Netflix documentary *My Octopus Teacher* is another example of the growing interest in encounters with beings radically different from the self. As research in anthrozoology—the study of human-animal relationships—delves further into the topic, it raises new questions about whether bonds with other species offer benefits that the bonds with fellow humans do not.

"What Does a Parrot Know About PTSD?"

In 2016, *The New York Times* published a piece called "What Does a Parrot Know About PTSD?" The article reported on unexpected bonds between human veterans and parrots at a bird rescue facility in Los Angeles called Serenity Park. Many of the veterans had severe post-traumatic stress disorder. Where conventional therapies had failed to treat their symptoms, these avian partnerships succeeded. "I see the trauma, the mutual trauma that I suffered and that these birds have suffered," observed one veteran while stroking a caique parrot named Cashew, "and my heart just wants to go out and nurture and feed and take care of them, and doing that helps me deal with my trauma. All without words."

Meanwhile, a nine-year-old diagnosed with autism spectrum disorder struggled in social interactions with her classmates and siblings. But in a study involving socializing with cats and dogs, she demonstrated a remarkable level of empathy and intersubjective understanding. Not only did she anticipate specific events related to their well-being ("But they'll get stuck!" she exclaimed upon learning that kittens can climb trees), but she also considered their subjective experiences of these events ("But they are afraid!"). Clinical psychologist Olga Solomon researches the unique benefits children with autism derive from interacting with companion animals, challenging the predominant story of the disorder as an impairment in social abilities.

And at a specialized care facility for individuals with Alzheimer's disease, scientists placed fish aquariums in the activity and dining area. When compared with control groups, patients in the presence of these aquariums had improved outcomes (including increased nutritional intake and body weight), adding to growing evidence that animal-assisted care alleviates tension, isolation, and helplessness felt by individuals suffering from chronic illnesses.

Until quite recently, the therapeutic use of animals received little attention from mainstream clinical literature and public funding initiatives. Before considering why this might be, it's important to note that there are potential stressors and challenges for the animals involved in these therapies, requiring strict ethical guidelines that safeguard the welfare of both patients and therapy animals throughout the treatment process. Some animal-assisted therapy programs instrumentalize other animals as mere means toward human ends—another instance of human exceptionalism. However, perhaps the limited attention these interventions have received *reflects* this pervasive bias more than anything. In a 2010 piece about animal-assisted therapy, for instance, the veterinary scholar James Serpell wrote, "It is difficult to escape the conclusion that the current inability or unwillingness of the medical establishment to address this topic seriously is a legacy of the same anthropocentrism that has dominated European and Western thinking."

Today, animal-assisted interventions are gaining recognition in a variety of therapeutic settings, including general hospitals, palliative and critical care units, nursing homes, and school special-needs programs. They have shown promise in treating trauma, depression, anxiety, and various other medical conditions, including chronic pain, fibromyalgia, obesity, eating disorders, and cardiovascular diseases.

The benefits of close contact with other animals extend well beyond formal therapeutic settings. In the United States alone, people live with nearly four hundred million companion animals (encompassing more than two-thirds of all households), and more than 90 percent of those asked report regarding their animals as family members. Studies consistently show that living with other animals not only buffers against stress and loneliness but also confers a sense of identity, self-worth, and existential meaning.

Meanwhile, nature-based interventions are on the rise. This trend includes ecotherapy and "green prescriptions," where medical professionals recommend nature activities—such as gardening and hiking—as part of a treatment plan for various physical and mental health conditions. Research has famously shown that the mere presence of shrubs, trees, and flowers outside a patient's room significantly accelerates the healing process. Horticultural programs have also yielded promising results in correctional contexts; formerly incarcerated individuals who completed their community service in outdoor or horticultural environments show lower rates of recidivism than those who completed their community service in indoor, nonhorticultural environments.

For years psychological research has shown that relationships are the foundation of a meaningful and healthy life. As a social psychology student, I learned that humans have a "need to belong"—an innate motivation to form and maintain interpersonal bonds with other people. According to Roy Baumeister (one of the theory's founders), belonging is a need, rather than a desire, because those who can't satisfy it suffer various mental and physical health deficits. But why limit this need to belong to human-to-human relationships? That didn't make sense to me. Like many people, I have always felt a profound sense of belonging amid other species—one that transcends words and is steeped in shared instincts, experiences, emotions, and histories. What if we took a humbler view of the potential contribution of other species to our well-being?

There is compelling evidence that nurturing connections with other species enhances our human relationships to boot. For instance, while most research emphasizes the "one-to-one benefits" of bonds with companion animals, studies also reveal a "ripple effect" whereby companion animals facilitate human social interactions and community. In a similar vein, community garden programs not only promote individual wellness but also contribute to a sense of neighborhood stability, stronger civic engagement, and a reduction in youth crime.

Knowledge that contact with other animals and natural environments is good for people is an ancient belief that research continually gives new weight to. But importantly, it implies serious consequences for our well-being as society becomes further estranged from the natural world (and of course, this is to say nothing of the vital roles other species play in maintaining healthy, functioning ecosystems).

I and Thou

During the COVID-19 lockdowns, virtual sessions of my Harvard class The Arrogant Ape took place over Zoom. Behind the confines of our screens, a small group of students and I met weekly to discuss assigned materials—readings, podcasts, occasional artworks—crafted by and for other humans. The irony of this approach (discussing human projects with other humans through human-made technologies) in a course that advocated for more-than-human interaction was not lost on me. In a small effort to counter this pattern, I introduced a weekly "slow looking" exercise, a practice I continue to implement in class today. Slow looking is based on the idea that if we really want to *get to know* something or someone, we must invest time in observing them. The exercise prompts students to dedicate a more generous part of their day than usual—whether ten minutes, an hour, or an entire afternoon—attending to a subject of their choosing (though the term "slow looking" uses the vernacular of the visual, observation can occur through any of the senses). Many students opt for familiar choices like a companion animal, houseplant, or nearby tree, while some go for more unconventional choices (spending time with a rock formation or a river).

The first time I introduced the exercise, I made the mistake of telling students their slow-looking experiences would form the basis of their final paper. Yet creating the expectation that one should get something concrete out of the exercise defeats its entire purpose. Slow looking is not a directive on how to observe or what, if anything, to extract from the chosen subject; instead, it encourages patient, immersive attention that can bring you into the present moment and foster a deeper connection with that subject.

In his 1923 book *I and Thou*, the philosopher Martin Buber draws a relevant distinction between what he calls I-It and I-Thou ways of seeing. In I-It relationships, the other is an "it" that exists only as an instrument or means to an end—something to be analyzed, categorized, and used for specific purposes. In contrast, I-Thou relationships involve openness and complete presence. In an I-Thou encounter, I neither objectify nor "interpret" the other but honor them as a unique, irreducible, and valuable being. Importantly, this "other" need not be human; Buber famously gives the example of different ways of relating to a tree. Though both ways of seeing are part of human experience, Buber emphasizes the importance of nurturing I-Thou connections to promote authentic relationships and a more profound, meaningful life.

Buber's I-Thou mode of relating aligns with the encounter I described at the outset of this chapter—the one where I suddenly found myself alone for the first time with a wild baboon. She and I were relating as two subjects, embodied minds at work. In that moment, there was a reciprocal recognition of each other's existence that transcended any need for interpretation or projection. And through that sense of shared presence, the agency and consciousness of the other appeared self-evident. Perhaps being one-on-one as our peers slept helped, allowing us to shed aspects of our respective species identities and ways of being. It was at this point that I began to grasp the essence of phenomenologists' point that mind largely manifests in and as relationship, rather than as some "objective fact" in the world awaiting discovery or verification.

In Buber's terms, the mind of the other is an *experience* afforded by an I-Thou encounter, not something readily measurable or quanti-fiable through an I-It connection. This is perhaps what the immersive observations of Barbara Smuts (with baboons) and Len Howard (with birds) had in common. These researchers each blurred the sup-posed boundary between observer and observed that typifies the sci-entific gaze. Through mutual relationship with their nonhuman collaborators, they allowed the animals to speak for themselves. The I-It mode that distances the self from the other (ultimately rendering the other into an object of manipulation) has implications far be-yond science. Too often, therein lies Nature's value—in its instru-mental utility, not in the inherent worth of a jaguar, a forest, or a waterfall. This losing logic is driving the ecological crisis we face to-day. Accumulating more and more data on other beings and the en-vironment doesn't appear to be enough: this crisis is not something to solve "out there," but is a crisis of our relationship with the world.

In Buber's famous adage, "All real living is meeting."

And as we'll see next, like all beings, human beings don't just *have* relationships—they *are* relationships.

CHAPTER EIGHT

The Entangled Bank

TAKE A LONG, DEEP BREATH IN. NOW SLOWLY LET IT OUT. Each time you inhale, you're drawing in oxygen from the plants around you. Once in your lungs, oxygen navigates the bloodstream, where it gets exchanged for carbon dioxide. With each exhale, you fill the air with carbon dioxide, the very substance that all these plants need for photosynthesis.

What we breathe out, plants breathe in. What plants breathe out, we breathe in. The air you breathe is the collective breath of other living beings.

You are immersed in the world.

And the world is immersed in you.

Your body hosts a remarkable diversity of life; as many as one thousand different species dwell on your skin, in your mouth, and in your gut. Only about 10 percent of your cells carry the human genome, while the remaining 90 percent harbor genomes from bacteria, viruses, fungi, and other microorganisms. This multispecies collective (also known as the microbiome) keeps you alive—it facilitates digestion, metabolism, immunity, neurological function, and other vital processes.

Delving deeper, seven *octillion* atoms exist in your body, each billions of years old, forged in the core of an ancestral star before eventually becoming part of you. That is, perhaps, what the naturalist John Muir meant in observing that "when we try to pick out anything by itself, we find it hitched to everything else in the universe."

So when we try to dominate, we actually disrupt the cycles on which we ourselves depend. It's the ultimate act of hubris.

How are cutting-edge scientific insights revealing striking interdependencies among different species, including between humans and other forms of life? Traditionally, how has anthropocentrism fueled an essentially competitive, hierarchical view of Nature? How does this in turn obscure our understanding of evolution and of ourselves in ways that perpetuate notions of human exceptionalism? How has this emphasis on competition led people to misinterpret and misuse evolutionary theory to explain the current anthropogenic crisis of life on earth? How can a less anthropocentric understanding help us to reimagine it?

"Nature, Red in Tooth and Claw"

It's feeding time at Chimfunshi, a sanctuary for chimpanzees in the Copperbelt region of northern Zambia. Chimfunshi is one of the oldest and largest chimpanzee sanctuaries in the world. It covers a vast area of forested land where 150 chimpanzees, many victims of the illegal bushmeat and pet trade, now roam. The chimps sleep in nests and spend much of their time exploring the outdoor wooded areas, emerging several times a day for supplementary food from human caretakers.

I stand on the observation deck as eager pant-hoots fill the air. Dozens of hunched-over, shadowy figures begin materializing from the brush for their afternoon snack. The chimpanzees soon congregate and pace below the deck where caretakers are preparing sugarcane, sweet potatoes, and cabbage. In these high-density, high-stakes circumstances, the potential for conflict in the group is elevated. The social tension becomes palpable as the chimpanzees exchange vigilant glances and vocalize with increasing intensity. Nobody seems at ease. Yet what surprises me is the absence of aggression and hostility amid it all. Instead, some of the chimps start engaging in reassurance behaviors. They approach one another and embrace, body-kiss, or put their finger or hand in another chimp's mouth (a sign of trust). Jake Brooker, one of my student collaborators, has also observed an increase in sociosexual interactions during such socially tense moments. Often, it's the individuals you may least expect who readily participate in these comforting behaviors, such as two burly, high-ranking males. This is not to say there is no conflict or competition; indeed conflicts do sometimes break out. But this potential for competition and conflict is what facilitates cooperative acts like reassurance.

The way we see evolution shapes the way we see ourselves.

Think back to when you first learned about evolution. If you're like me, phrases such as "survival of the fittest" and "struggle for existence" come to mind. These terms tend to evoke a competitive, selfish model of Nature. This view—sometimes called "nature, red in tooth and claw"—depicts organisms engaging in a perpetual battle for resources, territory, and dominance.

However, most scientists today would agree that most major events in the history of life on earth were also the result of enormous cooperation and symbiosis. As noted above, mutualistic relationships abound among microbes, fungi, plants, and animals like us.

Cooperation is neither the antithesis of conflict nor some rare exception in evolution. So why does this stereotype of "nature, red in tooth and claw" persist in the public and scientific imagination?

The emergence of evolutionary theory in the mid-nineteenth century coincided with the rise of industrial capitalism. Darwin's central ideas were thus interpreted in ways that resonated with the competitive ethos of this growing capitalist system. Social Darwinism emerged later that century to apply evolutionary ideas like natural selection to human societies. It posited that societal progress was driven by competition: those who excelled in the competitive market were regarded as more evolutionarily successful and inherently superior. Similarly, poverty and failure were attributed to the supposed inferiority of certain individuals or groups. As one might imagine, this perspective provided a pseudoscientific rationale for existing social hierarchies and economic disparities.

For instance, it was not Charles Darwin but Herbert Spencer, an influential English sociologist and proponent of Social Darwinism, who coined the phrase "survival of the fittest." In an effort to connect his racist economic theories with Darwin's biological principles, Spencer posited that social hierarchy was not only justifiable but also reflective of the most advanced and resilient societies. Darwin himself was more cautious about applying his own theories directly to human society. Nevertheless, his ideas on the mechanisms of natural selection in evolution offered a seemingly natural and scientific justification for capitalist and imperialist narratives of competition and the pursuit of self-interest.

The individualistic competitive worldview is reflected in other popular metaphors, such as the "selfish gene," put forward by the evolutionary biologist Richard Dawkins. As metaphors gain popularity, we unfortunately tend to lose sight of the fact that they are mere analogies. Dawkins himself has repeatedly clarified that selfish genes don't necessarily make for selfish individuals. On the contrary, selfish genes can lead to all kinds of altruistic behavior in individuals!

Darwin also stressed that natural selection is not a process by which organisms independently vie for supremacy. For instance, upon introducing the term "struggle for existence" in *On the Origin of Species*, he explains, "I should premise that I use the term Struggle for Existence in a large and metaphorical sense, *including dependence of one being on another*" (emphasis added). In the book's famous final paragraphs, Darwin invokes an entangled bank—filled with many species of plants, birds, worms, and insects—to illustrate this interdependence. Years later, he would suggest in *The Descent of Man* that sympathy is a fundamental evolutionary force in social animals: "It will have been increased through natural selection; for those communities, which included the greatest number of the most sympathetic members, would flourish best and rear the greatest number of offspring."

In short, the co-optation of evolution into a purely competitive and individualistic framework offered a narrow and often distorted view of Darwin's theory, one mirroring the broader societal trends and ideologies of the time. But even during that period, various scholars issued strong challenges to this one-sided view. One notable rebuttal came from the Russian anarchist Peter Kropotkin in his widely read 1902 book, *Mutual Aid: A Factor of Evolution*. Kropotkin asserted that cooperation is abundant in Nature and plays a vital role in the overall well-being of individuals and societies. "Don't compete! . . . Practice mutual aid! That is what Nature teaches us," he exclaims. "That is the surest means for giving to each and to all the greatest safety, the best guarantee of existence and progress, bodily, intellectual, and moral." Kropotkin proposed that the modern emphasis on competition was anthropocentric, likely a reflection of our own strivings and failings rather than of how Nature works.

Is Nature fundamentally competitive or cooperative? A lot seems to hinge on how we answer this question.

But why must we choose one? Animals like chimpanzees are no more inherently violent and competitive than they are peaceful and cooperative. As described earlier, reassurance behaviors multiply when the potential for conflict is highest, revealing how cooperation and competition themselves are entangled. One begets the other. Competitive problems often require cooperative solutions. Those who cooperate better typically compete better. Life requires the management of both competitive and cooperative relationships.

Just think of the people you have the most conflict with. Next, think of the people you cooperate the most with. If you're like me, the answers are the same. Our most involved and intimate relationships—whether with partners, family members, close friends, or colleagues—often demonstrate the entangled nature of competition and cooperation.

Yet popular literature and media often question whether human nature is essentially cooperative *or* competitive, a dichotomy exemplified by the stark contrasts drawn between bonobos and chimpanzees, humans' closest living relatives. Those favoring a peaceful view of human nature tend to endorse bonobos' female-bonded "make love not war" reputation, while those leaning more toward the "nature, red in tooth and claw" outlook emphasize the stereotype of the male-dominated, aggressive chimpanzee. My collaborators and I have shown, however, that the social behavior of these two ape species is more similar than often assumed. Through an analysis of various chimpanzee and bonobo sanctuary communities, we've found that variation *within* the two species is greater than differences *between* them. For instance, consider empathetic responses like consolation and sociosexual interactions during consolatory acts. Based on existing stereotypes, one might reasonably expect such friendly behaviors to be more prevalent in bonobos than chimpanzees. But group differences reveal a far more nuanced pattern: some communities of chimpanzees look more bonobo-like, and some communities of bonobos look more chimpanzee-like.

So are we innately hostile and violent toward others (supposedly like chimpanzees) or friendly and peaceful (supposedly like bonobos)? A glance at the news will also suggest that the answer is not so straightforward: Humans possess the capacity for both aggression *and* cooperation. Shouldn't we afford the same recognition to our closest primate relatives, rather than categorizing them into rigid species stereotypes?

Competition and cooperation are both driving forces in evolution. The point is not to emphasize one over the other but to recognize the complex interplay between the two. But how has the conventional emphasis on competition influenced our scientific approach and, consequently, our understanding of Nature's deeper workings?

Subversive Science

Suzanne Simard, a professor of forest ecology at the University of British Columbia, grew up roaming Canada's old-growth forests with her family, exploring moss-covered trails, foraging for mushrooms, and building forts and rafts from fallen branches.

As a young student in forestry school, she learned an accepted dogma: life in the forest was governed by competition. According to this view, trees were solitary individuals constantly competing with one another for access to sunlight, water, and nutrients.

At the same time, Simard was growing concerned with the rise of commercial logging projects clear-cutting diverse forests and replacing them with homogenous, single-species plantations. In many ways, the conventional competitive view sanctioned these forestry practices, emphasizing techniques like weeding, spacing, and thinning to favor specific individuals or species. The idea behind these "free-to-grow" initiatives was that by reducing competition from other vegetation, the newly planted trees would thrive.

But compared with the trees in the old-growth forests Simard had come to know and love, these newly planted trees proved more susceptible to disease and climatic stress. Without competitors, they were less healthy. For instance, Simard noticed that when nearby trees like paper birch were removed, planted Douglas fir saplings were more likely to get sick and die. But why? The planted saplings had ample space and received even more light and water. Why did they fare noticeably worse?

Simard eventually obtained a grant to test her hunch that the answer was hidden beneath the soil. If planted seedings were mixed with other species, she hypothesized, they might survive better through some kind of underground support system involving their roots. To test this idea, Simard planted birch and fir trees together and traced how carbon molecules went back and forth between the two. Her groundbreaking doctoral research found that birch and fir trees collaborate by exchanging carbon through the underground fungal networks connecting their root systems (a.k.a. mycorrhizal networks). The more shade the birch trees cast on the fir trees, the more carbon was sent over to the fir. Essentially, there was a net transfer from birch to fir that compensated for this shading effect. Upending the long-held view that species were always competing, Simard's research was featured on the cover of *Nature* in 1997, which called these networks the "wood wide web."

Since then, Simard and her students have discovered extensive mycorrhizal networks connecting the trees within an area of a forest. They are often connected to one another through an older tree she calls a "mother" or "hub" tree who shares nutrients with other trees and young saplings. The fungal network helps not only with nutrient exchange but also in protecting the plants against pests and disease. However, there is another side to this coin. When plants are unable to carry out photosynthesis themselves, they may resort to extracting resources from others through these shared mycorrhizal networks. And not all chemicals moving through the networks benefit the receiving plant: for example, plants can also distribute toxic substances that hinder the development of neighboring plants.

Though Simard's research landed in a top scientific journal, she faced intense backlash and criticism for challenging conventional forestry science, a male-dominated field. As she recalls in a 2020 *New York Times* interview: "The old foresters were like, Why don't you just study growth and yield? . . . I was more interested in how these plants interact. They thought it was all very girlie." Skepticism about Simard's research persists, in part because of entrenched beliefs that humans are the only species capable of such elaborate cooperation. This skepticism is also fueled by the suggestion—frequently amplified by the media more than Simard herself—that trees *always* benefit from being connected by mycorrhizal networks. Such singular narratives overlook the variety and complexity of relationships possible in the forest. The forest is both a collaborative and competitive ecosystem. It's again about this intricate interplay, this give-and-take, this essential balance defining any living, evolving, dynamic relationship.

Simard explains her frustrations with the tendency of Western science to overlook these relationships. "We don't ask good questions about the interconnectedness of the forest, because we're all trained as reductionists. We pick it apart and study one process at a time, even though we know these processes don't happen in isolation."

As Simard acknowledges, this interconnected ecological perspective has long been part of many animist and Indigenous views of reality, which approach the world through relationships of reciprocity. As we'll see in the next chapter, today's cutting-edge Western scientific findings tend to agree much better with such worldviews than is commonly presumed. Yet even throughout Western history, numerous scientists have defied reductionism in favor of interconnection. Instead of regarding Nature as a collection of discrete objects, Darwin saw a densely entangled web of subjects. The revered German philosopher Goethe championed a holistic approach to studying the natural world, expressing that "in nature we never see anything isolated, but everything in connection with something else which is before it, beside it, under it, and over it." His friend the great naturalist Alexander von Humboldt similarly believed in studying relationships between different elements of the natural world rather than isolating them: "Everything," Humboldt wrote, "is interaction and reciprocal."

In the late nineteenth century, the development of ecology—the study of the relationships among living beings and their physical surroundings—offered a formal challenge to the principles of scientific reductionism. Ecology earned a nickname as "the subversive science" for its power to make humans reconsider their place in the natural world. A notable offshoot is deep ecology, conceived by Norwegian philosopher Arne Naess in the 1970s. This environmental philosophy explicitly rejects anthropocentrism, emphasizing the intrinsic value of all living beings and acknowledging the profound interconnectedness that defines our existence.

Fungi and trees are so interconnected that some scientists believe they should not be viewed as separate organisms; instead, the forest functions as an integrated entity. According to the principles of deep ecology, everyone is deeply entangled with everyone else. Humans are no exception. So then where does Nature end and do we begin?

"We Have Never Been Individuals"

Influential thinkers have cautioned that using terms like the "natural world" and the "environment" (as I've done for convenience in this book) risks suggesting that Nature lies somewhere beyond ourselves. That is, the very existence of a word and concept like "nature" reinforces a dualistic understanding of the natural world as distinct from human culture or society. The anthropologist Marilyn Strathern's studies of the Hagen people in Papua New Guinea exemplify how not all humans adhere to this so-called nature-culture divide. In a paper titled "No Nature, No Culture: The Hagen Case," she argues that unlike most Euro-Americans, the Hagen people have no absolute opposition between nature and culture through which they comprehend the world. In a similar vein, Strathern argues that the notion of an "individual" (from the Latin *individuus*, meaning "indivisible") as a separate and autonomous "self" is rooted in Western ideas of individualism. This notion views a person as a self-contained entity, entirely defined by the limits of their own body. In Melanesian societies, in contrast, individuals are not isolated units but primarily defined by their position and relations within a social network. She proposes the word "dividual" to characterize the relational nature of personhood in such societies.

How can updated knowledge of biological relationships among living beings also reframe our understanding of individuality? One fascinating example is the lichen. No matter where in the world you reside, you have probably encountered one. If you're in New England like me, think of those crusty sage-green formations you see adorning tree trunks and rock surfaces, though lichens come in myriad colors and forms. The plant-like appearance of many lichens, along with their ability to photosynthesize, led early naturalists to categorize lichens as a type of plant. It wasn't until the late nineteenth century that scientists discovered that lichens are actually collaborations between two organisms: a fungus and an alga. The fungus provides structural support, nutrient absorption, and water retention, while the alga contributes through photosynthesis, supplying essential energy to the fungus. The partnership allows lichens to thrive in diverse environments, from the harsh Arctic tundra to the most arid desert landscapes. A lichen is not a *singular entity* but a *composite being.*

Lichens led the German botanist Albert Frank to coin the term "symbiosis" in the late 1870s. Symbiosis refers to close, long-term physical associations between members of different species (when the association benefits all parties, it's a particular kind of symbiosis called mutualism). Since the term was introduced, symbiosis has been found to play an essential role in the development and survival of almost every organism. It is a ubiquitous feature of life.

Consider the algae that power coral reefs. Years ago, I was snorkeling on the Great Barrier Reef and noticed patches of coral reef bleaching. I had assumed that elevated ocean temperatures (due to global warming) caused these once colorful and thriving coral formations to fade. It turns out that corals have a symbiotic relationship with microscopic algae living in their tissues. When water is too warm, corals expel the algae, leading to a loss of nutrients and pigmentation, making the corals appear white. So it's not that rising ocean temperatures are bleaching the corals per se, but rather that they are disrupting the *relationship* between coral reefs and their algal symbionts.

Partnerships between plant roots and mycorrhizal fungi, as mentioned earlier, form another one of earth's most widespread symbioses. We also have symbiosis to thank for the mitochondria that make our cells run. Mitochondria originated from a free-living bacterium that got swallowed up by an ancestral bacterial host some 1.5 billion years ago. But instead of being digested, the bacterium formed a mutually beneficial relationship with the host, providing energy in return for a protected environment and nutrients. The process came to be known as endosymbiosis.

Endosymbiotic theory, first proposed by the evolutionary biologist Lynn Margulis in the late 1960s, explained the presence of mitochondria in our cells (and chloroplasts in plant cells, which were thought to originate from a similar endosymbiotic event). It showed that complex lifeforms, including animals, plants, and fungi, evolved from simpler, symbiotic relationships. Margulis's theory pushed back against the prevailing scientific emphasis on competition at the time:

"The view of evolution as a chronic bloody competition among individuals and species, a popular distortion of Darwin's notion of 'survival of the fittest,' dissolves before a new view of continual cooperation, strong interaction, and mutual dependence among life forms. Life did not take over the globe by combat, but by networking. Life forms multiplied and complexified by co-opting others, not just by killing them."

Nature is not a zero-sum game, where one entity's gain is necessarily the other's loss. Yet like so many of the revolutionary thinkers we've encountered, Margulis was initially scoffed and laughed at by the scientific establishment. She was denounced as a scientific radical, apparently even critiqued for upending biology in favor of creationism (the equivalent of academic heresy). Her manuscript was rejected more than a dozen times before it was finally accepted. Today, endosymbiotic theory is the leading evolutionary theory for the origin of eukaryotic cells—those constituting our life and that of all complex organisms. It is considered one of the great discoveries of twentieth-century evolutionary biology. Not bad for a heretic!

Picture an evolutionary tree, with species diverging from one another over time, each on their own trajectory until they settle on separate branches. Lynn Margulis's endosymbiotic theory offers an alternative perspective, emphasizing how organisms readily interact and influence one another—more like a web or a net than a tree. Building on Margulis's insights, the anthropologists Carla Hustak and Natasha Myers propose a new term: "involution." Unlike the word "evolution" (which literally means "rolling outward"), "involution" suggests a "rolling, curling, turning inwards," where living beings continuously intertwine themselves in processes like symbiosis.

Perhaps even the image of an evolutionary tree reflects a cultural bias toward individualism—of atomized, competing individuals striving in parallel. We're neither standing atop a ladder nor perched at the tip of a twig. We're enmeshed in a wide and deep net of symbiotic relations.

Because we coevolved with plants, for instance, we often experience a pleasant sensation when we eat them. Imagine savoring a deliciously ripe blueberry. What a clever strategy on the part of plants—to bear fruit with such delectable flavors, enticing animals like us to eat them so we then spread their seeds. This long coevolutionary partnership has led to a diversity of fruit types and tastes, with different plant species adapting to the habits of specific animals. For instance, avocado plants, with their large fruit pits, originally evolved alongside megafauna such as mammoths, horses, and giant ground sloths—animals sizable enough to disperse their seeds. Our eyes, too, are adapted to perceive the vibrant colors of fruits and flowers, helping us animals easily spot ripe, edible plants in the environment. How enriching it is every time I come to recognize and experience one of these coevolutionary processes. Our bodies and senses have evolved in delicate reciprocity with the lifeforms surrounding us. We cannot separate humans from other beings—indeed we *are* who we are because of them. As my friend the cultural ecologist David Abram puts it, "We are human only in contact, and conviviality, with what is not human."

Developments in the microbial sciences have also made it hard to define the boundaries of an individual organism. It's no longer possible to think of "you" as distinct from the microbial communities you share a body with. You are one big symbiont, what researchers have called a "holobiont" (from the Greek *holos*, meaning "whole"; *bios*, "life"; and *ont*, "to be"), an ecosystem in and of yourself.

By cell count, the vast majority of what you might consider "your" body is not actually yours—it contains trillions of microorganisms, outnumbering your human cells by ten to one. The number of bacteria in your gut alone exceeds the number of stars in our galaxy. The number in your mouth is comparable to the total number of human beings who have ever lived on earth! If one were to remove all these microbes from the body and put them on a scale, they'd weigh in at about three pounds—the same weight as an average human brain. And research suggests they can wield as much influence as the brain. Your ability to solve complex memory and learning tasks is predicted by the health of your gut flora. Your mood, too, depends in part on the composition of your gut bacteria (as suggested by the colloquial "gut feeling"). For instance, interventions that alter the gut microbiome (such as probiotics) have shown promise in regulating behavior and brain chemistry associated with depression and anxiety.

The immune system also develops in close dialogue with your microbiota. At any given moment, these unseen partners are helping mediate your response to other organisms. They shape not only how you fight disease but also how you digest and derive nutrients from the environment. Microbes extend the capabilities of their hosts, who rely on this symbiotic relationship for their very existence. For instance, cows themselves can't eat grass, but their microbial populations can. Over time, animals and their microbial partners have co-evolved so closely that unique bacterial strains are adapted to a particular animal niche. As one example, 90 percent of the bacterial species residing in termite guts are not found anywhere else in the world (importantly, this also means that for every animal species who goes extinct, some unknown number of highly specialized bacterial lineages also disappear).

All these findings trouble the idea of a discrete, autonomous entity known as "the self." Our microbiome is dynamically shaping who we are in ways we are only beginning to understand. Of course, not all aspects of this relationship are harmonious. There are many situations where the interests of the symbionts don't align. For example, a bacterial species in our gut may be essential for digestion but could also lead to a fatal infection if it enters our bloodstream.

In late 2019, humanity encountered a new microbe—SARS-CoV-2. I remember walking through Harvard's deserted campus months after the COVID-19 lockdowns began. The quad and library steps, usually bustling with students and conversation, sat strangely silent. A half-finished coffee had been left untouched on my desk since March. Notices plastered on bulletin boards served as reminders of the new reality, urging adherence to safety protocols and social distancing measures. My partner remarked how a microscopic organism originating on the other side of the globe had brought one of the world's most respected establishments to its knees. It was humbling and surreal to think of it this way. Through events that were (and remain today) largely mysterious to us, an invisible pathogen in the form of a virus had upended our social lives, norms, and institutions.

Viruses like SARS-CoV-2 use human bodies as tools to reproduce, as do many bacteria (albeit in their own manner). I love to reflect on the philosopher John Gray's point that we are technological devices invented by ancient bacterial communities as a strategy for genetic survival. Under this view, microbes, rather than humans, are the driving forces of innovation. Myra Hird, professor of environmental studies at Queen's University, elaborates on this idea: "Most organisms are bacteria: they evince the greatest organismal diversity, and have dominated evolutionary history. Bacteria invented all major forms of metabolism, multicellularity, nanotechnology, metallurgy, sensory and locomotive apparatuses (such as the wheel), reproductive strategies and community organization, light detection, alcohol, gas and mineral conversion, hypersex, and death."

The virus demonstrated how a tiny microbe can outmaneuver and profoundly impact the health and behavior of individuals and populations. COVID-19 is believed by many to have zoonotic origins, meaning it was transmitted to humans from other animals. With humans coming into more and more contact with other animals—whether through factory farms or habitat encroachment—epidemiologists warn that widespread illness and pandemics are more likely.

If we held greater respect for Nature—the power she has not only to be harmed by us but to harm us in turn—perhaps we would see more clearly our interdependence. Perhaps we would try harder to protect other species if we understood that in doing so we are also protecting ourselves. If we became more aware of this complex whole we inhabit, perhaps we would live in a way that supports planetary well-being.

In 2012, a team of respected biologists published a paper titled "A Symbiotic View of Life: We Have Never Been Individuals." In it, they draw on recent technological advances and scientific discoveries, like those highlighted in this chapter, to argue that it is high time we rethink the notion of a "biological individual" in favor of a recognition of interspecies interdependences. The article concludes with a bold declaration: "For animals, as well as plants, there have never been individuals. This new paradigm for biology asks new questions and seeks new relationships among the different living entities on Earth. We are all lichens."

The Year 3011

I recently came across a cartoon by the artist Dan Piraro titled "The Year 3011." The cartoon depicts two ants, clad in togas, sitting amid the remains of ancient Greek pillars and temples—pondering over the ruins of human civilization. A callout bubble shows one ant asking the other: "And yet, can a species that eliminates itself in just a few million years be called 'successful'?"

Despite our apparent evolutionary "success" as a species, it's likely that other lifeforms—among them ants, lichens, and countless others—will endure long after humans' tenure on earth. Science fiction novels (such as those that inspired *Planet of the Apes*) imagine a future earth run by other species in the aftermath of humanity's self-destruction. If given the opportunity, would these other forms of life come to dominate the planet to the extent that human activities have?

As highlighted earlier, evolution isn't just about ruthless competition; the history of life on earth is equally marked by widespread cooperation and symbiosis. Yet despite this evidence, prominent thinkers today continue to promote the identification of evolutionary "success" with dominance over the rest of Nature. A recent *Scientific American* article titled "What Makes Humans Different Than Any Other Species" exemplifies this perspective:

"Why are humans so successful as a species? [Humans and chimpanzees] share almost 99 percent of their genetic material. Why, then, did humans come to populate virtually every corner of the planet—building the Eiffel Tower, Boeing 747s and H-bombs along the way?"

A brief aside: I would not cite nuclear weapons as evidence of our species' "success."

However, perhaps the article is merely acknowledging humans' remarkable capacity to manipulate and control their environments. But even in this aspect we are not without rivals. Just take cyanobacteria—some of the earliest photosynthesizing organisms—responsible for the rapid oxygenation of earth's atmosphere during an episode known as the Great Oxidation Event. Billions of years ago, they set the conditions for life as we know it today, causing the extinction of many anaerobic organisms (those not requiring oxygen) and allowing aerobic lifeforms (those requiring oxygen) such as animals, plants, and fungi to evolve and thrive.

Zoologist Luis Villazon explains for the BBC that even humans' claim to ecological dominance represents a narrow view:

"Humans have certainly had a profound effect on their environment, but our current claim to dominance is based on criteria that we have chosen ourselves. Ants outnumber us, trees outlive us, fungi outweigh us. Bacteria win on all of these counts at once. They existed four billion years before us, and created the oxygen in the atmosphere. Collectively, bacteria outnumber us a thousand, billion, billion to one, and their total mass exceeds the combined mass of all animals."

Measuring and defining evolutionary success by a particular kind of dominance in which humans happen to excel, let alone dominance at all, is a self-serving perspective. One can also see why this characterization is human-centric via examples of species who are successful by other means.

Mosses provide a helpful illustration. As Robin Wall Kimmerer has shown, mosses have thrived on this earth for more than three hundred million years (compared with *Homo sapiens'* meager 200,000), thanks to very low competitive ability. These tiniest of plants survive by collaboration—building soil, purifying water, and creating a viable home for many other forest creatures. What if cooperation were the means by which evolutionary "success" was measured and achieved? Or qualities like longevity, resilience, and the ability to sustain thriving interspecies communities?

But humans sit at the top of the food chain—isn't that evidence of a natural hierarchy? A food chain offers a simplistic, linear view. A more realistic representation of consumption relationships in ecosystems is food webs, which consist of many interconnected food chains, where organisms at different levels mutually influence one another. Yet so long as we want to think in a linear fashion, plants are the top of the *producer* chain. They possess the miraculous ability to convert sunlight into food for animals like us. Without them, our existence would be inconceivable. Does this imply that plants are superior to humans? Then there are fungi, relishing their place atop the *decomposition* chain, recycling organic matter (such as dead plants and animals) into simpler compounds while promoting soil fertility, nutrient cycling, and the health of plant communities. Why establish hierarchies based solely on consumption—a value deeply embedded in capitalist culture—when Nature's relationships can be described in myriad ways?

The standard view of evolution emphasizes competition for resources. Human ascendancy over the natural world can seem like a logical, inevitable consequence of our own natural selection through this lens. Accordingly, the ecological crisis sometimes gets framed as an inevitable part of the evolutionary process: the logical outcome of humans acting in their own self-interest. In a similar vein, scholars and journalists often claim that the human mind is simply not designed to solve the problem of climate change. There are evolved psychological barriers, so this story goes, that prevent us from acting to address it on the scale that is required. Let's call it the inevitability narrative. You can probably tell from my tone that I don't agree much with it.

For one, as Quentin Atkinson and my colleague Jennifer Jacquet have argued, the inevitability narrative disregards profound variation within and between human cultures in the way people respond to climate change. There is no universal human response to this issue. Framing climate inaction as part of human nature (by suggesting it's not only natural but inevitable) is a way to justify the status quo. It also conveniently frames responsibility for climate change in terms of the individual rather than cultural values, norms, and institutions (including corporate actors).

Sometimes, the inability of other species to survive in human-altered environments and the resulting mass extinction event underway get portrayed as the normal progression of "survival of the fittest," an unfortunate artifact of humans' evolutionary success. However, it's essential to recognize that such extinction events defy the parameters of natural selection. Natural selection is evolution through differential survival and reproduction. But when a species is under the extreme pressure that leads to extinction, there is no room for the differential survival and reproduction essential for evolution.

In sum, the ecological crisis we face today (and our current collective failure to address climate change on the scale required) is not some inevitable upshot of the human evolutionary process. This misinterpretation once more places undue emphasis on the role of competition and individualism. Instead, our legacy should be examined through a much broader lens that recognizes evolution as a process of continuous adaptation and acknowledges the powerful role of culture.

So what will be our legacy? In the year 3011, what will the ants be saying?

The God Species

Several years back, I attended a talk by the renowned scientist David Keith on solar geoengineering. Solar geoengineering aims to counteract global warming by reflecting sunlight away from earth's surface, usually by injecting reflective aerosols into the stratosphere. As I listened to the presentation, I became increasingly bewildered by its implications. The technology appeared eerie and outlandish—more the stuff of science fiction than academia. Yet as Keith argued, if it could be realized (and notably, recently, the first outdoor test in the United States took place), solar geoengineering could potentially slow, stop, or even reverse the rise in global temperatures in just a few years. So even as I resisted, I found myself wondering: Why haven't we yet taken the actions necessary to reduce global emissions and avert climate catastrophe? And perhaps more urgently, what is the alternative if we continue down this path of inaction?

I began feeling nostalgic for blue skies (solar geoengineering could result in a hazy, white appearance to the sky). Questions started to swirl in my mind regarding inadvertent consequences, such as the impact of dimming the sun on other species—including pollinators like honey bees—who rely on sunlight to navigate and find food (I have since learned that scant research exists on this question, despite our food system relying on answers). Furthermore, I wondered, might this intervention de-incentivize other efforts to reduce carbon emissions (a.k.a. the "moral hazard" of geoengineering)? Not to mention that the vast majority of scientists involved in solar geoengineering research hail from elite American and European universities, with growing concerns about the technology's unequal distribution of risks in rich and poor countries. But above all, I found myself grappling with the uncomfortable realization that solar geoengineering exacerbates human dominance over Nature precisely when we urgently need to curtail it. I kept asking myself, isn't there an inconsistency between the positive ecological values the use of these technologies purports to serve and the mindset these same technologies reinforce within our culture?

In her bestselling 2021 book, *Under a White Sky*, Pulitzer Prize–winning journalist Elizabeth Kolbert takes a hard look at human attempts to actively manage and control natural systems to address environmental challenges—engineering the atmosphere and oceans, manipulating genomes, electrifying rivers, assisting migrations, and introducing novel species to manage those deemed problematic. Kolbert reveals how even the most well-intentioned interventions often yield unintended consequences, inadvertently harming ecosystems and disrupting global weather patterns. This triggers a domino effect, leading to more complex problems that demand evermore inventive solutions. The more we attempt to defy Nature, the more obvious our own limitations become. And yet paradoxically, the very sorts of interventions that have imperiled our planet are increasingly seen as our only lifeline.

Darwin's entangled bank reminds us that human beings are just one species among many interconnected within the great web of life. In these intricate networks of cause and effect, it's no wonder that human interventions often yield unintended consequences! As ecologist Frank Egler highlights, "Nature is not only more complex than we think. It's more complex than we can think." As a result, human technology frequently struggles to reproduce the invaluable capacities of intact, healthy ecosystems.

For instance, when human pollutants threatened aquatic ecosystems, we engineered water treatment plants in their place. Yet despite significant investment, these facilities often fall short when compared with their natural counterparts. Natural ecosystems excel in filtration and biological degradation processes—not to mention feats like flood control and biodiversity support. So while water treatment plants are helpful, it's like trying to replace a whole orchestra with just one instrument.

Or consider the human-caused extinction of native pollinators in the United States, animals finely attuned through coevolution with native flora (such as squash bees, whose biology evolved to perfectly match the floral traits of squash flowers). When European settlers introduced honey bees to North America in the seventeenth century, these honey bees became crucial for agriculture. But they didn't do nearly as well. The novel environment made them more vulnerable to stressors like colony collapse disorder—which continues to imperil honey bee populations and the stability of our food system today.

And in most cases, as Suzanne Simard observed, our attempts to cultivate forests pale in comparison with Nature's prowess. Likewise, in industrialized agriculture, pesticides, fertilizers, and genetic modifications are employed to support the growth of single high-yield crops, often at the expense of the polycultures nurtured by natural ecosystems—resulting in less resilient and healthy agricultural landscapes. When we view life as a competition and position ourselves as its masters, we overlook the need to work with, rather than against, these naturally interdependent processes.

This doesn't mean that technological innovation has no part to play in addressing ecological degradation. However, I am convinced that we are not going to get very far with such interventions unless we first question human dominion and sovereignty over Nature. Many technofixes are deployed today in the name of saving the environment, yet they often reflect only the human exceptionalism that has driven its destruction in the first place. If we want to chart a truly sustainable course forward, we will need to address the root problem rather than its symptoms.

———

Another form of human exceptionalism, sometimes called human *exempt*ionalism, suggests that humans are so superior that they are immune to the environmental influences and constraints that govern other forms of life. Human exemptionalism has come to constitute a commonsense view that our species' distinct ingenuity and resulting technologies will allow us to transcend current ecological problems. This unwavering faith in progress and innovation endows humans with an almost divine ability to always know and do better (imagine an updated Great Chain of Being, where human-made technologies have replaced the angels and gods). Yet for all our attempts to tame and control Nature, we remain just as subject to ecological cycles and interdependencies as other species. That we see ourselves as exempt from Darwin's entangled bank is evident in our ambitions to both geoengineer earth and colonize distant worlds like Mars.

Biosphere 2 was an experimental facility in Arizona designed to simulate earth's ecosystems in a closed environment for scientific research, particularly to gather data for maintaining human life in outer space. The project encountered myriad challenges. Among them were problems growing trees. As it turns out, trees need wind. Without the movement of air to carry moisture away from their leaves, trees struggle with mold growth and reduced oxygen production. And when there is minimal wind (as is the case on Mars), there is little resistance to that wind, and the trees grow up weaker and less resilient. This major obstacle underscored the significance of considering the complex interactions between organisms and their environment.

Imagine a world of desolate vistas stretching endlessly under an orange-red sky. The day-night cycle, atmospheric conditions, and gravitational pull on planets like Mars stand in stark contrast to earth's familiar rhythms. The thin Martian air is composed primarily of carbon dioxide, rendering stepping outside without a full space suit impossible. The altered gravity, about 38 percent of earth's, would reshape your movements and weaken muscle strength and coordination. Your internal clock, attuned to earth's twenty-four-hour cycle, would struggle to adapt to the longer Martian day, relying on artificial lighting to regulate itself. Earth's vibrant ecosystems are conspicuously absent from Martian landscapes. There is no wildlife or vegetation to be found. No sign of microbial life. There are no oceans or streams, no wolves howling or cardinals singing, no scents of honeysuckle or fresh rain. You'd never feel the wind brush across your face. It's the absence of life as we know it.

Space colonization promises to carry the human enterprise onward and forward. But the realities of interstellar travel are far more sobering. It's very unlikely we'd survive the trip (beyond earth's protective magnetic field lie lethal levels of radiation), let alone the physical and psychological challenges of life on another planet. Despite what Elon Musk and Jeff Bezos want you to believe, we are born and bred of this earth, our bodies and minds evolved in deep reciprocity with the forces of *this* planet over millions of years of evolution. *To imagine exempting ourselves is to ignore the truth of our entangled existence.* The cautionary tale of Biosphere 2 serves as a stark reminder of the irreducible complexity of ecosystems and the potential for unforeseen environmental consequences. So while earth-exit strategies may sound innovative, I think they are better understood as new—and arguably more powerful—expressions of human exceptionalism.

Human dominance fosters a psychological detachment from the natural world, reducing other lifeforms (and even entire planets) to mere commodities for exploitation. Yet in this endeavor, we too ultimately suffer. The more we seek to manipulate and control Nature, the greater the harm we inflict upon ourselves—an insight powerfully elucidated by Rachel Carson in her 1962 book, *Silent Spring*, which exposed the detrimental effects of pesticides on the environment and human health. As Carson poignantly remarked, "But man is a part of nature, and his war against nature is inevitably a war against himself."

Even in Death

Finally, recognizing our entanglement with everything else may mean rethinking death itself. Anthropologist Deborah Bird Rose explains that ecological communities—relations between predators and prey, producers and consumers, parasites and hosts—rely on continual intergenerational cycles of life and death. She discusses the interconnectedness of these two processes within ecosystems, emphasizing that death is not merely an end but a vital part of the circle of life. It's all part of the same journey.

And yet increasingly—and related to the idea that we are exempt from the entangled bank—another unifying theme of Western culture is avoiding the reality of death by any means necessary. Immortality projects like cryopreservation are on the rise, with people opting to freeze their bodies (or often, just their heads) after death with the hope of future revival. Transhumanists envision a future where humans merge with advanced AI systems to extend the human lifespan indefinitely. Fantasies of uploading our minds to computers raise the tantalizing possibility that people could one day outlive their biological animal bodies. Under this view, our true essence is a disembodied consciousness—some kind of floating mind—rather than a whole feeling, moving, breathing, sensing animal being. Philosopher Derek Parfit summarized this perspective when he wrote, "The body below the neck is not an essential part of us."

In *How to Be Animal*, author Melanie Challenger argues that these ambitions are the heirs of ancient dualist ideas that separate the mind from the body and devalue our embodied, interdependent existence. She underscores how the minded life of a human being—what we might call our "experience"—is intimately affected not only by our brains but by everything from our gut bacteria to the state of our various limbs, organs, and environments. Yet our desire to perceive the human mind as something special leads to disdain for our animal bodies, a discomfort only heightened by our awareness of mortality. It fuels a quest to evade death through technology rather than accepting it as a natural part of the cycle of life. Ultimately, Challenger suggests that our perception of mental distinction becomes our hope for salvation.

In 1973, cultural anthropologist Ernest Becker published his Pulitzer Prize–winning book, *The Denial of Death*, in which he outlined how fear of our own mortality leads us to deny our nature as animals. Becker's examination of how humans psychologically cope with mortality laid the groundwork for Terror Management Theory, which proposes that when confronted with our mortality, we tend to distance ourselves from other animals because they remind us of our own mortal existence. In support of this, numerous studies show that reminders of our vulnerability to death trigger a strong psychological need to proclaim that "I am not an animal."

For instance, one 2001 study presented American college students with open-ended questions concerning thoughts and feelings about their own death—such as "Jot down, as specifically as you can, what you think will happen to you as you physically die and once you are physically dead." Participants then read an essay that either described humans as distinct from other animals or emphasized the similarity between them. When prompted to consider their mortality, participants showed a clear preference for the essay emphasizing human uniqueness over the essay on human-animal similarity.

A later study tested this phenomenon in a consumer marketing context. A 2010 Wrangler jeans advertising campaign titled "We Are Animals" pictures people imitating different animal behaviors (e.g., a woman wearing jeans crouching to the ground to drink water in a forested area or people running through red smoke in different animalistic postures). In the study, participants viewed either the original advert or a modified ad that included the same images but instead read "We Are Not Animals." When existential anxiety was high (measured via the accessibility of death-related thoughts), consumers favored the brand image promoting the modified "We Are Not Animals" message over the original advert. Such findings support Terror Management Theory's contention that thoughts of mortality lead us to deny our own animality. And yet it's our mortality that sustains and nourishes other life, making death integral to the vitality of eco-systems. If we recognize this inherent interdependence, do we have as much to fear?

⸻

One of the most amazing things I've ever seen is the *Welwitschia mirabilis* plant, endemic to the Namib Desert in Namibia and southern Angola. Some of these plants are estimated to be more than two thousand years old. Then there's Old Tjikko, the nickname given to a Norway spruce tree in Sweden, estimated by carbon-dating to be more than 9,500 years old. Many plants have the ability to continuously regenerate or clone themselves and thus, at least in a genetic sense, could be considered immortal. Lichens can be successfully resuscitated after ten years of dehydration. Wood frogs can remain frozen for up to eight months in the winter—their hearts stop and they cease respiration—before returning to life in the spring. And *Turritopsis dohrnii*, also known as the immortal jellyfish, possess the extraordinary capacity to revert their cells back to a younger state, essentially reversing the aging process. Their ability to potentially achieve biological immortality has captured the interest of scientists studying aging and regenerative medicine.

Intelligent life is not something we need to seek elsewhere: it abounds in diverse and truly awe-inspiring forms right here on planet earth—a paradise of shapes, textures, habits, feelings, minds. And all of us are entangled, creating an irreducibly complex net of relations. What a brilliant place to call home and partake in. How much escapes us when we presume to comprehend it entirely or attempt to bend it to our will.

The carbon building blocks of your body came from animals, plants, and other beings who lived millions of years ago. Perhaps this is why my former biology teacher liked to say, "I'm 3.7 billion years old" (referring to the estimated emergence of life on earth; my physicist friend instead claims he's 13.8 billion years old, the approximate age of the universe). It's through perpetual cycles of life and death that we find ourselves here today. Resisting this exchange and attempting to dominate and control Nature is futile, for, as anthropologist Tim Ingold explains, "One cannot appropriate that within which one's being is wholly contained."

We talk about *our* oceans, *our* planet, but could we reframe this possessive language to instead reflect that shared sense of *belonging*? To a place that co-constitutes our very being? Perhaps then we'd no longer be quite so afraid of death. By embracing our entanglement with the natural world—by realizing we have never just been individuals—we may find solace in the knowledge that death is not an end but a continuation of life. The recent rise in "green burials" and human composting suggests we are already moving in this direction. For me, unlearning human exceptionalism brought a noticeable decrease in fearing death. It's not that I don't fear it, but there is undeniable beauty in the idea of decomposing by bacteria and fungi, returning nutrients to the soil, becoming an ancestor to beings yet to come.

In *Learning to Die: Wisdom in the Age of Climate Crisis*, Robert Bringhurst and Jan Zwicky argue that embracing our mortality and recognizing the interconnectedness of all life on earth can enhance our understanding of our role in the world and inspire more meaningful, humble action in addressing environmental crises. We will always have a bit of the wild in us, they write, "because we are part of it—and cannot, even in death, be disconnected from it completely."

CHAPTER NINE

Our Indigenous Inheritance

I N 1994, A GROUP OF CAVE SCIENTISTS IN SOUTHEASTERN
France stumbled upon a mound of fallen stones that appeared to
block a passageway. What they discovered behind it is now considered among the most famous prehistoric rock-art sites in the world.
Chauvet Cave, like other Paleolithic caves, contains a curious puzzle
that has long intrigued scientists and artists. No human figures are
depicted on its walls; only masterful illustrations of other animals—
over a dozen species including lions, hyenas, horses, deer, bears, and
rhinos. According to many scholars, this is suggestive evidence that
our hunter-gatherer ancestors did not perceive the world anthropo-
centrically, instead imbuing other animals with great wisdom and
spiritual power. As paleoarchaeologist Jean Clottes writes: "The essential role played by animals evidently explains the small number of
representations of human beings. In the Paleolithic world, humans
were not at the center of the stage."

What might a culture not based on the myth of human exceptionalism look like in practice? How have Indigenous communities
around the world employed and protected this non-anthropocentric

way of relating to and understanding Nature? How can Indigenous wisdom help point the way toward the kind of science that does not fall victim to human exceptionalism's biases? What evidence do we have that this humbler relationship in practice promotes biodiversity and ecosystem health?

Traditional Ecological Knowledge

I should first acknowledge that I cannot presume to fully grasp Indigenous knowledge. It comes from a way of knowing the earth that's different from what I experienced in my own upbringing and culture. It is also important to attend to the ways in which Indigenous knowledge—often termed "local ecological knowledge" or "traditional ecological knowledge"—can be misunderstood, simplified, or instrumentalized. Even speaking of "Indigenous knowledge" in general obscures an incredible diversity of beliefs, languages, values, customs, and historical experiences.

However, a common perspective shared by many past and present Indigenous cultures is that humans are deeply enmeshed within their environments. "Indigenous" comes from the Latin word *indigena*, meaning "sprung from the land" or "native." Indigenous knowledge is rooted in direct, lived experiences with the land refined over generations through relationships with other forms and ways of life. In the words of Tewa scholar Gregory Cajete, Indigenous knowledge emphasizes "mutual reciprocity, which simply means a give-and-take relationship with the natural world, and which presupposes a responsibility to care for, sustain, and respect the rights of other living things, plants, animals, and the place in which one lives." Rather than positioning humans as superior to or separate from Nature, such traditions recognize humans as part of a larger interconnected web of life. As Cajete elaborates, "Everything is related, that is, connected in dynamic, interactive, and mutually reciprocal relationships."

Another feature shared by numerous Indigenous cosmologies is the belief that agency and intelligence pervade the natural world. Myriad beings—not just animals and plants but also mountains and rocks, rivers and lakes, the sky and celestial bodies, the wind and weather—are thought to be imbued with spirit or consciousness. This perspective, as highlighted by the Native American theologian George "Tink" Tinker, stands in sharp contrast to Western commonsense knowledge and scientific convention:

"The Western world, long rooted in the evidential objectivity of science, distinguishes at least popularly between things that are alive and things that are inert, between the animate and the inanimate. Among those things that are alive, in turn, there is a consistent distinguishing between plants and animals and between human consciousness and the rest of existence in the world. To the contrary, American Indian peoples understand that all life forms not only have consciousness, but also have qualities that are either poorly developed or entirely lacking in humans."

For years, wildlife biologists operated under the assumption that badgers and coyotes living in the same territory solely competed for prey. However, upon closer observation, researchers realized a surprising truth: these two species often collaborate strategically during hunts, enhancing their overall success. The realization is more consistent with the worldview of many Indigenous peoples, who emphasize intelligence throughout Nature and reciprocal relationships between entities.

To explore this idea explicitly, bethany ojalehto and colleagues ran a study in which they removed all the text from a storybook on coyote and badger hunting. They then presented solely the illustrations to U.S. college students and Indigenous Panamanian Ngöbe adults. While the majority of American participants interpreted the story as competitive, most Ngöbe participants interpreted it as cooperative. Moreover, Ngöbe participants were much more likely than Americans to agree that animals like coyotes and badgers are capable of engaging in intentional communication and behaving morally. Many Ngöbe participants spoke of conversations, customs, and even religiosity among animals, endowing them with sophisticated mental and social capacities.

Anthropologists have observed that the Indigenous cultures they study often attribute a viewpoint to all beings, not just humans. This is evident in Eduardo Viveiros de Castro's notion of Amerindian "perspectivism," where "the world is inhabited by different sorts of subjects or persons, human and nonhuman, which apprehend reality from distinct points of view." Unlike Western cultural relativism, in which there are various cultural viewpoints on a shared natural world (many cultures, one nature), perspectivism involves a singular "human" point of view taken by a range of beings with distinct natures (one culture, many natures). As he summarizes, "Animals are people, or see themselves as persons." What this means is that other animals perceive the world in a manner akin to how humans do, but what they perceive differs depending on their specific corporeal natures. He famously uses the example of how what humans see as blood, jaguars see as maize beer. This is not to impose a "human" quality onto a jaguar, but to consider how it manifests from the jaguar's distinct point of view.

Eduardo Kohn's ethnographic work with the Upper Amazonian Runa likewise reveals a cosmology in which "all beings, and not just humans, engage with the world and with each other as selves—that is, as beings that have a point of view." Historically, the term "animism" was used in a pejorative and racist manner to describe such belief systems as primitive. However, religious studies scholar Graham Harvey's exploration of the animistic beliefs and practices of Native Americans, Māori, Aboriginal Australians, and eco-pagans has helped recast the term in a more sophisticated, accurate light: "Animists are people who recognize that the world is full of persons, only some of whom are human, and that life is lived in relationship with others." In animist worldviews, personhood is implicated in the very condition of being alive and in relation with others. Consciousness, in many such views, is not an "add-on" to life but rather its very essence.

However, beliefs in the pervasiveness of personhood and consciousness—the root of many Indigenous worldviews—have consequences. According to a customary Inuit saying, "The greatest peril of life lies in the fact that human food consists entirely of souls." Anthropologist Colin Scott's research with the Cree hunters of northern Quebec highlights how the killing of another being poses a moral dilemma that must be addressed through rituals of reciprocity, gratitude, and forgiveness. For the Cree, as for many Indigenous cultures, these rituals recognize the inherent worth of nonhuman beings and seek to compromise with them in ways that promote the sustainability of human and nonhuman life and the relations in between. But in an anthropocentric culture where other animals, plants, rocks, and entire oceans are viewed as mere resources, human consumption is often justified as an expression of our inherent superiority or God-given dominion.

Robin Wall Kimmerer, a member of the Citizen Potawatomi Nation, shows how these divergent cultural perspectives manifest in language. The Potawatomi language, like many Indigenous languages, is verb-rich, reflecting a worldview that recognizes the animacy of Nature. English, on the other hand, is a noun-based language (which she wittily deems "somehow appropriate to a culture so obsessed with things"). Whereas roughly 30 percent of English words are verbs, that proportion is 70 percent in Potawatomi. Nouns in English are often verbs in Potawatomi, suggesting that what English speakers perceive as static objects are understood as more dynamic and alive in her native language. For example, rather than using "fox" as a noun, the Potawatomi language might describe a fox with a verb that conveys the action of being a fox.

These linguistic conventions shape how speakers engage with the world, portraying other beings as active subjects rather than passive objects. According to Kimmerer, the limits of the English language are exacerbated by the conventions of Western science: "Science is a language of distance which reduces a being to its working parts, the language of objects." Nonetheless, she emphasizes that Western science and Indigenous knowledge each offer valuable insights, and when combined, can provide a more comprehensive understanding of the natural world.

Native Science

There is a growing movement to debunk the myth of human exceptionalism in Western cultural and scientific thought, one I am very much aligned with. However, haven't Indigenous ways of knowing—which challenge separations between "nature" and "culture" and "animal" and "human"—been articulating these concepts all along? Exploring these nondualistic perspectives was mind-opening and affirming, but it also made me wonder: Why had I hardly heard of any of this before?

Indigenous knowledge has frequently been marginalized or erased as pseudoscience. The dismissal reflects a broader colonial history of oppression faced by Indigenous peoples globally. Indigenous cultures have been and often remain in tension with dominant cultures. Genocide and forced assimilation have been used as tools not only to dispossess Indigenous peoples from their lands and resources but to systematically devalue Indigenous knowledge systems.

Yet many of today's "cutting-edge" scientific findings on animal and plant cognition confirm what these cultures have long known. For example, while Jane Goodall usually gets credited with "discovering" chimpanzee tool use and cooperative hunting, ethnographic records reveal that Indigenous African communities knew about these behaviors and reported them to European colonists hundreds of years prior.

Likewise, forest ecologist Suzanne Simard admits that she did not "discover" mycorrhizal networks. The Coast Salish people of the Pacific Northwest (where she studies old-growth forests) have long taught that trees are persons involved in an elaborate network of symbiotic relationships under the forest floor. Simard relates an account from Bruce "Subiyay" Miller of the Skokomish Nation, who said that under the soil, "there is an intricate and vast system of roots and fungi that keeps the forest strong."

Métis scholar Zoe Todd recounts a formative career moment while attending a talk by the French philosopher Bruno Latour. Latour's lecture understood the planet as an animated entity comprising reciprocal interactions among all forms of life (an entity he broadly refers to as "Gaia," drawing from Greek mythology). However, Todd noted with dismay that in discussing Gaia, Latour did not reference any Indigenous thinkers or ideas, including the relevant Inuit concept of "Sila," which similarly understands earth as an interdependent, intelligent, living entity. The intent is not to point fingers at Latour—a personal hero of Todd's—but to highlight a broader issue: the tendency of Euro-American scholars to overlook Indigenous contributions.

The rejection of human exceptionalism in Western scholarship is a movement sometimes called "posthumanism." Many posthumanist "realizations"—for instance, that nonhuman beings are sentient and possess agency—are deep-rooted in Indigenous beliefs and practices. When Euro-American thinkers get credited but not their Indigenous contemporaries who are addressing the very same topics in their work, it perpetuates the marginalization of Indigenous peoples and knowledge. It's a narrow lens that I am certainly guilty of. As just one example, the first paper on animal ethics I ever published begins with the line: "The moral status of animals is a longstanding question dating back at least to Aristotelian philosophy." But of course, inquiries into how humans should treat other animals go back *much* further than the Western philosophical tradition.

Unfortunately, many people raised in a Western cultural context are conditioned to believe that only one philosophy and one science exist. Gregory Cajete thoughtfully challenges this assumption in his 2000 book, *Native Science*:

"Some moderns, both scientists and non-scientists, argue that there is no such thing as Indigenous science, that science is essentially a Western construct or concept. And that while Indigenous peoples have folkways and folk knowledge, this knowledge is not scientific. The argument states that the term 'Indigenous science' is essentially meaningless. Others perceive science as a way of understanding the world, a story of how things happen, a way that human beings have evolved to try and explain and understand existence in time and space and relationships vis-à-vis the natural processes of the world. In this perspective, every culture has science."

Blackfoot scholar Leroy Little Bear similarly describes science as the pursuit of knowledge at the boundaries of human understanding. To the extent that this is true, he argues, there are certainly "sciences" beyond Western science, including Indigenous sciences. Ojibwe scholar Megan Bang and her colleagues highlight that even within Western science, there is a diversity of methodologies, theoretical underpinnings, and values. After all, they note, the United States has the National Academy of Scienc*es*, not the National Academy of Science. Currently, this organization has fifty-four separate sections, including over a dozen divisions focused on different aspects of biology alone!

When I began exploring Indigenous knowledge, I realized that much of what I had come to understand about "science" needed to be reevaluated. Many of us are taught to think of science as settled fact, rather than the great, flawed, ongoing process of knowledge construction that it is. For years, I accepted certain ideas without question, including that science was value-free—a series of cut-and-dried eternal truths. I believed that science was entirely divorced from the culture in which it operates. I've now come to see that one of the greatest myths within Western science is that it is immune to cultural influence.

I had a small lesson in this recently when my student collaborator Jake Brooker noticed something out of the ordinary at Chimfunshi, the chimpanzee sanctuary in Zambia where we conduct research. Two chimpanzee males, Max and David, were engaged in oral sex during a post-conflict encounter. Same-sex sexual behavior remains largely overlooked in chimpanzees, in part because it's more commonly associated with our bonobo cousins. But in a larger study, my colleague the primatologist Rachna Reddy found that sociosexual behavior in wild chimpanzees also occurs frequently between same-sex partners, particularly among adult male chimps.

In reality, same-sex sexual behavior has been recorded in more than 1,500 animal species. Its ubiquity in Nature has long puzzled biologists, many of whom consider same-sex sexual behavior an evolutionary conundrum, given its presumed fitness costs (resulting in no offspring). Recent work challenges this presumption by providing alternative hypotheses for how same-sex sexual behavior evolved, including highlighting its adaptive role in maintaining social relationships and mitigating conflict (as suggested by Jake's chimpanzee observation). A 2024 survey study among animal scientists found that same-sex sexual behavior is widely observed yet seldomly reported: most respondents (77.6 percent) had observed same-sex sexual behavior in their study species, but less than half (48.2 percent) collected data on it, and even fewer (18.5 percent) had published papers on it. Queer and feminist scholars have shown that evolutionary biology prioritizes behaviors deemed culturally acceptable while overlooking others, reinforcing ideas about what is real and "natural." As they've argued, the long-standing focus on heterosexual relations in science reflects patriarchal Euro-American cultural norms rather than the full spectrum of animal behavior.

Our beliefs and perceptions of what is "normal" filter what we observe in that full range of possibilities. At any given moment, our attention captures a mere slice of reality. This is also true for scientific knowledge at any given time. We often overlook that we live in a specific scientific era; what was once accepted as scientific fact has since been disproved (such as homosexuality being an evolutionary aberration). Western science is continuously changing, culturally situated, and value-laden—traits that are shared by all sciences.

Unfortunately, some people still disregard local ecological knowledge as subjective and arbitrary; Western science, in contrast, is seen as objective and rigorous. Others may be willing to admit that Indigenous cultures hold substantial knowledge of the environment but consider it valid only when "verified" by Western science. Both perspectives undermine the rich empirical foundations of Indigenous sciences, which are based on generations of careful observation, adaptation, and interaction with the natural world.

The meticulous scrutiny that Paleolithic cave painters devoted to their animal subjects is exemplified by the precision of anatomical features, postures, and behavioral repertoires featured. Research has shown that these ancient artworks often portray animal behavior, such as movement, more realistically than many of today's masterpieces. A recent study, for instance, found that prehistoric humans depicted the gait of four-legged animals much more accurately than modern artists. However, it's by the flicker of torchlight that Paleolithic cave paintings truly come alive. The animals seem to move across the walls like abstract figures in a surrealist dream. After a visit to the famous Lascaux caves in the Dordogne region of southwestern France, Picasso reportedly said to his guide, "They've invented everything."

The deep attention and intimate contact many Indigenous cultures maintain with Nature led anthropologist Louis Liebenberg to famously conclude that animal tracking was the original science. His research on the hunting techniques of the Kalahari San reveals how knowledge of animal behavior enables trackers to predict the actions of prey across specific times and environments. These trackers employ inductive and deductive reasoning to construct testable hypotheses about animals' whereabouts and revise those hypotheses as new evidence arises.

Tracking also requires deep empathy and intersubjectivity, often resulting in blurred identities between human and animal. As Liebenberg describes, "Tracking involves intense concentration resulting in a subjective experience of projecting oneself into the animal. . . . When tracking an animal, one attempts to think like an animal in order to predict where it's going. Looking at its tracks, one visualizes the motion of the animal and feels the motion in one's own body." Merging both somatically and mentally, the tracker effectively "must *become* that animal." This aspect of tracking aligns with a key principle of contemporary Indigenous science articulated by Gregory Cajete, who notes that Indigenous science is grounded in subjective experience, and the "perception gained from using the entire body of our senses in direct participation with the natural world."

Two-Eyed Seeing

Many of the core principles of Indigenous knowledge—recognizing other beings' autonomy and unique perspectives, emphasizing interdependence and cooperation, embracing embodied and empathic experiences—reflect ways of knowing Nature I've promoted in this book. As I've learned about Indigenous sciences, I've realized that these seemingly novel concepts in Western science have deep historical roots.

In 2023 I was invited to participate in a panel discussion with Sicelo Mbatha, an Indigenous philosopher and spiritual nature guide. The event, titled "Friendship with the More-Than-Human-World," explored how to relate to and understand other animals. Mbatha spoke of his close encounters with crocodiles, lions, and elephants, drawing upon extensive knowledge of the South African wilderness and traditional Zulu wisdom and practices. I shared stories about my work with wild baboons and other primates as a Western-trained scientist. Initially, I worried our different backgrounds and experiences with Nature might limit our conversation. However, Mbatha's deep connection with the land and my evolving scientific understanding found common ground. We both emphasized decentering human perspectives to study and relate to other animals on their own terms. We highlighted the value of immersive, empathic animal observations, and above all the importance of exercising humility when learning from Nature. At the end of the conversation, Mbatha said that he hadn't expected to find so much alignment with a Harvard scientist. I still consider it one of the greatest compliments of my career.

Vine Deloria Jr., a prominent Standing Rock Sioux scholar and activist, considered Indigenous knowledge the "intellectual twin to science." He underscored that Indigenous ways of understanding the world are rigorous, systematic, and deeply insightful—akin to what is recognized as science in Western cultures. These practices have been successful and sustained for many generations. Are they perfect? Probably not. Can Western science offer them insights? Almost certainly. Do they have lessons for Western science? Undoubtedly.

According to Robin Wall Kimmerer, bringing these two ways of knowing together does not mean *blending* them. Blending leaves you with neither of the original elements. Instead, she proposes knowledge symbiosis—a weaving together of different knowledges that each contribute equally. Mi'kmaq Elder Albert Marshall introduced the relevant concept of Etuaptmumk, or Two-Eyed Seeing, which involves "learning to see from one eye with the strengths of Indigenous knowledges and ways of knowing, and from the other eye with the strengths of Western knowledges and ways of knowing." Two-Eyed Seeing is an apt metaphor because it emphasizes the value of multiple perspectives. Each of our eyes captures a unique perspective because of its slightly different position on our face, a disparity allowing us to perceive the world in three dimensions.

In practice, Two-Eyed Seeing often involves collaborative research, education, and decision-making efforts between Western researchers and Indigenous communities. This approach shows great promise in conservation—including in understanding the behavior of ecosystems and wildlife, promoting biodiversity and sustainable land-use practices, and mitigating climate change. For instance, right now in Aotearoa, or New Zealand, a team of Western scientists and Indigenous scholars are weaving together modern genomic data with Māori knowledge practices to successfully translocate threatened freshwater fish and invertebrates. In the Canadian Arctic, Inuit knowledge of sea ice, wildlife migrations, and climate change has proved vital in regional conservation efforts. As just one example, the Kivalliq Wildlife Board is integrating Elders' and hunters' insights with scientific modeling techniques, a partnership facilitating better monitoring and protection of Arctic ecosystems and species like polar bears and ringed seals. And in regions like the Amazon basin,

Indigenous communities' extensive knowledge of ecosystems, medicinal plants, and sustainable agriculture is now aiding scientists in developing innovative fire-management and forest-renewal techniques.

More scientists today are embracing Indigenous knowledge in their work, which in no way diminishes the scientific contribution of that work. On the contrary, there's compelling evidence that it provides a more holistic and nuanced perspective of the animals, plants, and ecosystems they study. And by ascribing agency, intention, and communication to entities that Western science typically deems inanimate (including the land, elements, and stars), Indigenous knowledge can open the way to an expanded set of scientific principles and a deeper appreciation of previously overlooked ecological interactions.

At the same time, significant challenges remain in these cross-cultural collaborations. Though many Western scientists rely on Indigenous peoples to guide their research, these relationships have often been extractive and colonial. Scientists may draw heavily upon Indigenous knowledge without centering Indigenous struggles, compensating Indigenous communities, or including Indigenous people in the research (or do so merely as a token gesture). And importantly, despite Two-Eyed Seeing's call to value differing perspectives, there remains a tendency in the scientific community to assimilate traditional ecological knowledge within Western worldviews of Nature.

Place-Thought

Euro-Western frameworks often struggle to engage with Indigenous perspectives seriously. A major reason is because these cultures tend to approach knowledge very differently. Haudenosaunee and Anishinaabe scholar Vanessa Watts explains that Indigenous cosmologies often do not separate the nature of reality (ontology) from how one comes to know it (epistemology). She calls this concept Place-Thought, because *thinking* is indistinguishable from *being* in a specific place—a place with its own agency, thought, and will. *Thoughts* are shaped by the land and the animal, plant, mineral, and spirit worlds particular to a *place*. Consequently, knowledge is always local and relational: "Place-Thought is the non-distinctive space where place and thought were never separated because they never could or can be separated. Place-Thought is based on the premise that land is alive and thinking and that humans and nonhumans derive agency through the extensions of these thoughts."

Western commonsense knowledge rejects this notion, contending that meaning does not lie in the relational context of one's engagement with the world. Instead, it posits that meaning is imposed onto the world by the (human) mind. Thoughts, perception, and action are not derived from the land; they are separate from the supposed inertia of Nature. A mind detached from the world constructs a view of that world—it mentally *represents* the world. Indigenous ways of apprehending reality are not a matter of mental representation but of direct engagement—"not of making a view *of* the world but of taking up a view *in* it," anthropologist Tim Ingold explains.

The Western mind remains skeptical, rife with doubts, or may find itself squarely confused (if so, rest assured, you're in good company; grappling with this complexity is par for the course). Let me share a brief anecdote to illustrate the development of my own understanding of Place-Thought.

Sometimes, a changed worldview has a way of sneaking up on you. A few years ago, I attended a lecture by the environmental historian Bathsheba Demuth that had exactly that effect. Demuth detailed the hunting behavior of the Yupik people, an Indigenous group living along the Bering Sea coast who for centuries has relied on bowhead whales (bowhead whale meat and blubber remain integral to human sustenance in much of this region—even a very small bowhead can feed a village for more than six months). Elders pass down knowledge to younger hunters that human beings are part of an animate community of life that includes whales. In the Yupik worldview, humans are not inherently superior; rather, their existence is intertwined with bowhead whales, who also act as ethical agents—capable of judging and shaping how humans organize themselves in this shared society.

On St. Lawrence Island, generations of Yupik whalers have reported observing a relevant behavior. Upon seeing a whaling boat on the water, a bowhead swims nearby but stays just out of harpoon range—sometimes for over an hour—appearing to carefully attend to the people in the boat. Eventually, the bowhead decides to swim off or surface near the side of the boat where the harpooner is positioned. The Yupik refer to this latter action as *angyi*, from the root *ang-*, meaning the act of giving. Their interpretation of the whale's behavior is that the whale is assessing the worthiness of the people in the boat and deciding whether to offer her life to ensure the survival of their community. After careful deliberation, the bowhead may or may not choose to give herself to her hunters, consenting to die through her movements, an act critical for the hunters to take the next step to kill the whale.

The knee-jerk reaction of a Western scientist or ethicist might be that obviously the whales do not consent to die. Taking the liberty to speak on behalf of whales, one might view this as a convenient interpretation by Yupik hunters to assuage their guilt. Indeed someone in the audience made that very point: "I don't buy it," she said, this "mythology" that whales consent to die. Demuth's response, which I found compelling, was that she was not asking us to buy anything. She was not making a general statement about what *whales do or do not do* or what animals *are or are not capable of.* She clarified that the Yupik observations are not transferable. The unique environment and culture (Place) led to a specific set of behaviors and understandings (Thought), which would be very different in our own, Western context. Furthermore, hunting ethics on St. Lawrence and around the Bering Strait emerge not simply from human discourse. *They come from people taking the behaviors of the world around them very seriously.* These observations are inherently place-specific; they arise from a historical set of local relations within a multispecies society that is different from our own.

According to Place-Thought, a "society" is guided by interspecies ethical structures and agreements. These interspecies responsibilities and obligations shape the social world, where "personhood" is not determined by species or by human-likeness. Instead, humans, like other beings, are considered "persons" when they act in ways that this multispecies society deems appropriate and respectful. In the Western view, thoughts, perception, and action are not derived from the land and its manifold inhabitants; they are separate from the supposed inertia of Nature. Separating constituents of the world from how the world is understood (what Watts calls the epistemology-ontology divide) in turn limits agency to humans and leads to an exclusionary relationship with Nature: "When an Indigenous cosmology is translated through a Euro-Western process, it necessitates a distinction between place and thought. The result of this distinction is a colonized interpretation of both place and thought, where land is simply dirt and thought is only possessed by humans."

Who am I—shaped by a dominant culture known for its intensive animal confinement, invasive animal research, relatively limited contact with Nature, and education system that often reinforces these norms—to define what's real or right in a relational context I've never experienced? Who are we to distance ourselves from the land and then question those who've lived intimately with it for centuries? How can we speak on behalf of bowhead whales if we've never known one?

Decolonizing thought requires great humility and openness to ideas that might seem outlandish. Dismissing Yupik understandings of the world as impossible, mythological, or superstitious mirrors the colonial mentality that historically denied Indigenous peoples their culture, language, and reason. As Watts articulates: "Our understandings of the world are often viewed as mythic by 'modern' society, while our stories are considered to be an alternative mode of understanding and interpretation rather than 'real' events. Colonization is not solely an attack on peoples and lands; rather, this attack is accomplished in part through purposeful and ignorant misrepresentations of Indigenous cosmologies."

Colonial regimes forcibly relocated people from their traditional lands. This not only disrupted livelihoods but fractured local ecological knowledge, which is formed in intimate relation with place. As Indigenous studies scholars Eve Tuck and K. Wayne Yang have argued, true decolonization therefore must involve the return of territories to Indigenous peoples, rather than merely symbolic changes (like including Indigenous knowledge in textbooks). Colonial structures restrict and reduce relationships with land to one we have come to call "property." Yet there is growing recognition that current legal and economic frameworks fail to protect Nature because they consider Nature fundamentally as something that can be owned. Fortunately, there are promising signs of change.

Rights of Nature

In 2008, Ecuador became the first country in the world to grant legal rights to Nature. The updated Ecuadorian Constitution states that Nature—or Pacha Mama, a term from Indigenous Andean cosmology referring to Mother Earth—has the right to exist and maintain vital cycles, functions, and evolutionary processes. One of the first notable cases under this new legal framework occurred three years later when the provincial court of Loja ruled in favor of the Vilcabamba River's rights, halting a government road project that threatened the river's health. This groundbreaking case set a precedent for defending Nature's rights in court.

Following Ecuador's lead, other countries around the world have considered and implemented similar legal frameworks. For example, in 2014, New Zealand granted legal personhood to the rainforest Te Urewera, a former national park and the spiritual homeland of the Māori *iwi* (tribe) known as the Tūhoe. The Te Urewera Act declared the rainforest as a legal entity with "all the rights, powers, duties, and liabilities of a legal person." Courts in Bolivia, Colombia, and India have also recently ruled that ecosystems possess rights, and the movement is now gaining traction in countries like the United States, Mexico, Pakistan, Turkey, Nepal, and Australia. Many of these developments have been accompanied by an acknowledgment of Indigenous sovereignty and understandings of Nature, representing a paradigm shift toward including traditional ecological knowledge and values in environmental governance.

The granting of rights to Nature, particularly through legal personhood, is a relatively new movement in Western legal traditions. It began in 1972, when law professor Christopher Stone published a seminal article—"Should Trees Have Standing?"—that explored the possibility of recognizing the legal rights of forests, oceans, mountains, and other natural entities. Otherwise, he claimed, those entities were valued only according to their worth to humans and could be exploited. Stone noted that throughout legal history, each successive extension of rights to some new entity—including women and enslaved peoples—was previously considered "a bit unthinkable." Therefore, a compelling case could be made for the seemingly unthinkable idea of extending legal rights to trees.

Western legal systems typically divide the world into two categories: "persons" and "things." Whereas "persons" are subjects of rights, "things" may be owned as property. Today, only humans are considered persons in the legal sense, with few, telling exceptions (e.g., corporations and ships may also be considered persons!). While there's no single agreed-upon definition of a "legal person," it generally refers to an entity that possesses both rights *and* obligations. The rights-of-nature movement runs into trouble with the latter idea when it comes to entities like rivers, forests, or Nature as a whole. Can we hold a river accountable for flooding? Or a forest for burning? Such questions have led some to believe we need to break away from legal concepts like personhood, which were never meant to apply to Nature in the first place.

Additionally, unresolved questions remain about whether granting legal personhood to Nature fundamentally shifts colonial and anthropocentric paradigms or merely repackages them. Stone acknowledged that rights-of-nature efforts still unfold within a legal system that is itself inherently anthropocentric, where humans retain the authority to *recognize* the rights of other forms of life. Yellowknives Dene professor Glen Coulthard criticizes the shortcomings of this approach, noting that "where 'recognition' is conceived as something that is ultimately 'granted' or 'accorded' a subaltern group or entity by a dominant group or entity," this "prefigures its failure to significantly modify, let alone transcend, the breadth of power at play in colonial relationships." In courtrooms worldwide, human beings are the agents negotiating the moral frameworks. Yet alternative, interspecies ethics exist—as exemplified by Place-Thought and Yupik interactions with bowhead whales—in which other beings are also given agency in establishing and negotiating moral relations.

_____

Notwithstanding these ongoing debates, many see the incorporation of Indigenous worldviews into legal (rights-of-nature) and scientific (Two-Eyed Seeing) frameworks as a crucial step toward a sustainable relationship with Nature. Indigenous peoples manage or have tenure rights over more than a quarter of the world's land surface. This amounts to more than thirty-eight million square kilometers across eighty-seven countries or politically distinct areas. Research reveals that annual deforestation rates inside Indigenous territories are two to three times lower than in comparable regions, resulting in greater carbon sequestration and improved climate resilience. It is estimated that Indigenous peoples safeguard 80 percent of the world's remaining biodiversity in the forests, deserts, grasslands, and marine environments in which they've lived for centuries. Notably, biodiversity levels on these Indigenous-managed lands often equal or exceed those in conventional protected areas. And despite tremendous pressures from industrial activities, biodiversity is declining less rapidly on Indigenous territories than in other ecosystems. Consider the plight of nonhuman primates: Of the 521 primate species, around 68 percent are threatened with extinction due to habitat loss from agriculture and other industrial activities. Spatial analyses indicate that Indigenous lands are home to 71 percent of all primate species. As primates' range on these Indigenous lands increases, they are less likely to be classified as threatened or have declining populations.

"Indigenous peoples sustain nature because we know we are a part of nature," summarizes Steve Nitah, managing director at Nature for Justice—an organization that empowers Indigenous-led conservation efforts. Human exceptionalism, like all cosmologies, has practical implications for how its adherents interact with the world around them. The trends above provide compelling evidence that a less anthropocentric worldview predicts healthier human-nature relations and more resilient environments.

And yet like those environments, Indigenous knowledge and ways of life continue to be under threat. The World Bank estimates there are nearly half a billion Indigenous peoples worldwide. Although they constitute 6 percent of the global population, they account for almost 20 percent of the extreme poor and have substantially lower life expectancies than non-Indigenous peoples. Indigenous communities often lack formal recognition of their lands and are underserved by public investments. They still face significant barriers to participating fully in dominant economic, political, and judicial systems, inequalities that only increase their vulnerability to ecological destruction. Today Indigenous languages are being lost at an alarming rate. Studies suggest that similar forces are driving biodiversity loss and cultural and linguistic homogenization, further underscoring why efforts to dismantle human exceptionalism must involve decolonization strategies.

"Remember to Remember"

Indigenous peoples tell us over and over again that not all cultures set themselves apart from or above Nature. They remind us that the dominant Western anthropocentric worldview and exploitation of the environment are neither inevitable nor "natural."

I am not suggesting that Indigenous communities are monolithic. Nor am I implying that all human cultures that have lived in intimate contact with the land have done so sustainably and harmoniously. Some anthropologists argue that the initial arrival of humans in various parts of the world coincided too closely with local mass extinctions to be mere coincidence. This correlation is the basis of the "overkill hypothesis," which posits that the spread of early human hunting led to the extinction of many large animals (megafauna) at the end of the Pleistocene epoch, nearly twelve thousand years ago. While some evidence supports this hypothesis, the causes of these extinctions remain hotly debated, with many scientists attributing them to environmental changes rather than to human activity. So whereas some article titles claim "Humans Caused Megafauna Extinction in Australia" and "Global Late Quaternary Megafauna Extinctions Linked to Humans, Not Climate Change," others insist the opposite: "Humans Did Not Drive Australia's Megafauna to Extinction, Climate Change Did" and "Climate Change, Not Human

Population Growth, Correlates with Late Quaternary Megafauna Declines in North America." At best, according to critics, the over-kill hypothesis is an oversimplification.

At worst, it becomes a rationalization. Some use the overkill hypothesis as an analog to modern-day human impacts, suggesting that human beings have "always" exploited their environments. *The Guardian*'s George Monbiot exemplifies this perspective when he writes, "Anyone apprised of the palaeolithic massacre of the African and Eurasian megafauna . . . must be able to see that the weapon of planetary mass destruction is not the current culture, but human-kind." Paul Martin, the main proponent of the overkill hypothesis, also endorses this pessimistic view in an influential paper aptly titled "40,000 Years of Extinction on the 'Planet of Doom.'" If humans have been causing mass extinctions for eons, the logic goes, then no wonder we continue to be the destructive environmental force we are today. Yet this perspective overlooks marked individual and cultural variation in human-environment relations, including Indigenous pop-ulations who have maintained biodiversity and large areas of rela-tively unmodified forest for millennia, and continue to do so today.

Widespread beliefs in the "naturalness" of the human impact are both dubious and dangerous. They work to reify the worldview of hu-man supremacy—that it's our identity to be rulers of the planet, rather than *one* among many ways of being human. Such fixed ideas about "human nature" hinder our ability to explore and imagine who we can become. As Heidegger warned, any presumption to know "funda-mentally what man is . . . hence can never ask who he may be."

⸺

In the Western tradition, humans are considered the most advanced and evolved lifeform. In many Indigenous worldviews, in contrast, humans are seen as relative newcomers to life on earth—"the younger brothers of Creation"—with the least experience and thus the most to learn from other species. This humble perspective contrasts sharply with our common self-identification as *Homo sapiens sapiens*—the wisest of the wise. Humans may well be intelligent, but what is considered intelligent in Westernized contexts differs markedly from what countless traditional and Indigenous communities consider as such.

In the broad sweep of human history, Western culture is among the youngest and most inexperienced. Yet many of its hallmarks—including beliefs in human exceptionalism—seem simply a "matter of fact." Cultural and economic systems (like capitalism) that center human interests appear so given. However, if we condense earth's 4.6-billion-year history into a 46-year timeline, humans have existed for only four hours, and the Industrial Revolution began just one minute ago. Despite our attachment to our current way of life, it is relatively new, and it's important to recognize that alternative, sustainable ways of living have existed and thrived for millennia.

This is not to suggest that we go "backward," especially as Indigenous cultures and their knowledge are very much alive today. This is also not a revival of the "noble savage" argument, which idealizes Indigenous peoples in ways that are essentially racist. We should be able to ask what lessons we can learn from our species' rich and diverse history, acknowledging both successes and mistakes. And we could respect and learn from Indigenous knowledge systems without fetishizing or appropriating them. Yet Indigenous communities today are understandably cautious about the misappropriation of their cultural concepts and ideas. As just one example, plants are now being taken from the Amazon and patented by pharmaceutical companies without compensating the local Indigenous peoples who have contributed vital medicinal knowledge about them.

Dystopian and standard postapocalyptic narratives, such as those depicted in Hollywood films, often frame "the end of the world" as the collapse of modern civilization. However, Indigenous groups have been experiencing their own apocalypses for decades. Indigenous scholar Kyle Whyte suggests that if we seek resources for dealing with climate change and mass extinction, we might turn toward Indigenous communities, who have long endured these crises. Whyte argues that climate destabilization is just the latest assault of settler colonialism, noting that some Indigenous peoples "already inhabit what our ancestors would have likely characterized as a dystopian future."

We have much to learn from how Indigenous peoples navigate crises, relate to other animals and plants, engage with land, and practice science. Moving past human exceptionalism requires critical lessons contained in Indigenous ways of knowing that recognize the complex forms of agency, mind, and relationships in the natural world. But in truth, the recognition of the personhood of animals and plants is not confined to any one tradition. Non-anthropocentric ideas and practices can be found throughout Western history as well. For instance, panpsychism—the belief that mind exists in all things—is one of the oldest and most enduring Western philosophical concepts. From mystics and theologians to phenomenologists and feminist scholars, and from artists and poets to amateur naturalists and unorthodox scientists, various strands of Western thought have emphasized interdependence and kinship with the living world. While these themes have been more thoroughly explored in non-Western thought—especially in Indigenous and Eastern traditions where holistic perspectives and relationality are central—they are not entirely absent from the Western canon. They have merely been drowned out by louder voices. Human exceptionalism has long been the dogma of a dominant minority.

In *Native Science*, Gregory Cajete draws on Western scholars to argue that animism is a more universal human experience than often imagined:

"The word 'animism' perpetuates a modern prejudice, a disdain, and a projection of inferiority toward the worldview of Indigenous peoples. But if, as the French phenomenologist Merleau-Ponty contends, perception at its most elemental expression in the human body is based on participation with our surroundings, then it can be said that 'animism' is a basic human trait common to both Indigenous and modern sensibilities. Indeed, all humans are animists."

When I discuss posthumanism and the wisdom contained in Indigenous sciences with others, I'm often struck by how many people seem to intuitively "get it." As if these ideas are tapping into a much older and deeper way of knowing and moving in the world. Again, overcoming human exceptionalism is less about learning something new than about unlearning and shedding a dominant, culturally ingrained perspective. Humans, as Indigenous thinkers often note, do not need to invent an entirely new worldview to live by. The life-sustaining worldview that evolved with us is embedded deep within us, though often obscured. We are all indigenous to earth, native to this planet. Reawakening this inherent connection and unlearning human exceptionalism is possible because this knowledge already resides somewhere within us. All we must do, as Robin Wall Kimmerer relates, is "remember to remember." What's at stake are not only lessons to navigate the ecological crisis but also a more accurate and constructive account of how we became—and what it means to be—human.

CHAPTER TEN

Coming Down to Earth

NOTHING HUMBLES YOU QUITE LIKE BEING CHARGED BY A four-hundred-pound silverback mountain gorilla. His name was Munyinya and he lived on the slopes of Mount Sabyinyo, an inactive volcano straddling the Rwandan, Congolese, and Ugandan borders.

I was twenty-three years old and about to begin graduate school in New York City. I had encountered live gorillas only once before, at the Bronx Zoo. Now here I was, moments away from a face-to-face encounter in the wild.

A small group of us slogged up the mountainside through dense, thorny vegetation. The understory was so thick that even with experienced guides machete-clearing the way, we could see only a few feet ahead. Fire ants and stinging nettles competed for our ankles. We stopped every so often while one of us gasped for breath in the thin (8,500-foot altitude) air.

Finally we came to a clearing. Munyinya was resting under a thicket of bamboo surrounded by his harem. He sat peacefully like a big Buddha, legs crossed with hands folded gently in his lap. Two adult females munched on shoots while their young ones hung from and played among the vines (the smallest were swinging like it was their second language—somewhat awkwardly, and with effort). Our group sat down in a semicircle about twenty feet away. Every so often, Munyinya would look up at us. As instructed, we avoided making direct eye contact, something gorillas perceive as a threat. But when Munyinya produced a throat-clearing type of sound (*ummm-hmmm*), the guides replied in kind. Gorillas (and humans who speak some gorilla) use these vocalizations to convey a sense of contentment between individuals. All was well.

Then a photographer among us asked the guides to move a dangling branch that was obstructing his camera's focus. A sudden whack of the machete caused an unexpected ruckus. A huge bough fell to the ground, startling everyone, including Munyinya.

Before I knew it, Munyinya was lunging toward us. With fur raised, he was pounding his chest, roaring with mouth agape, vigorously shaking nearby vines to magnify the overall effect (as if being three times our size was not enough).

Munyinya paused mere feet away from us. In the heat of the moment my instincts told me to run, but a seasoned guide behind me forced my shoulders down. I slumped back to the ground, legs scratched and muddied—bipedalism suddenly proved useless. I stole a quick glance at Munyinya, who appeared irritated and irate. This was not the face-to-face encounter I had anticipated. I held my breath as Munyinya surveyed our group until gradually, inexplicably, the tension began to lift. After a series of intense back-and-forth grunts with the guides, Munyinya returned to his spot, sprawling on his back with hands clasped behind his head—as if to demonstrate his true, easygoing nature.

I'll never forget the way a juvenile on the sidelines was staring at me. She was still breathing heavily with eyes wide open, darting back and forth between us and her father. Her look communicated that we had just broken some cardinal rule and narrowly escaped retribution. Of course, she was right. Fortunately, the "laws" of Nature can be more forgiving than the ones we've manufactured. After fifteen minutes or so, Munyinya and his family brushed past us and disappeared into the bamboo forest.

There is clear wisdom in the back-and-forth of Nature—its conflicts, impasses, and desire for resolution. There is a remarkable balance; a power to ground you, to remind you of your place in the grand scheme of things. I followed gorillas into the jungle and suddenly found myself on their terms. I felt small, overwhelmed by Munyinya's ability to overpower us, to completely control the situation. It was humbling. It was terrifying. It was magnificent.

This awe-inducing experience has stayed with me for years. Psychologist Dacher Keltner and his colleagues describe awe as an emotional response to vastness that transcends one's current understanding of the world. Vastness is not merely about literal size—it can refer to capability, complexity, or any phenomenon that requires new ideas to accommodate what is being perceived. As Keltner explains, something is vast if it "is experienced as being much larger than the self, or the self's ordinary level of experience or frame of reference." Perhaps that's why we think of childhood as such an enchanted time—not due to some naive ignorance, but because *everything* feels vast, demanding constant accommodation and evoking wonder. Charles Darwin likened awe to a disorienting and enlightening sensation: "The mind is a chaos of delight"—a phrase he famously used upon first glimpsing a true tropical rainforest soon after the *Beagle* reached Brazil in 1832. Like Darwin's awe-inspired accounts of the natural world, many people consider experiences in Nature to be the most common elicitor of awe.

Awe has recently captured the attention of scientists in part because it is linked to various beneficial outcomes. These include a collective mindset—feelings of oneness, interconnectedness, and belonging—along with heightened prosocial behavior. Awe also seems to keep egotism in check, prompting a reduction in self-focus and making personal concerns and goals appear less significant. These effects are partly explained by what researchers call "the small self"—a relatively diminished sense of self in light of something deemed vaster than the individual. Awe's ability to locate us in forces larger than ourselves is one reason why it's not always enjoyable (as reflected in the word "awful"). One can be awestruck by a storm, a spiritual epiphany, or a gorilla charge, stirring admiration and wonder but also fear and uncertainty. At its best, awe expands our understanding of the cosmos and our place in it, much like the experience I had with Munyinya that day.

Too often, human exceptionalism thwarts our capacity for awe, blinding us to the beauty, power, and mystery of everyday existence. It fosters self-illusions that make us less open to our own weaknesses and limitations (the foundation of the human superiority complex), preventing us from appreciating the gifts other beings bring to the world and to our lives. In contrast, experiences of awe shift self-perception in the opposite direction, leading an individual to see themselves more accurately and to fully appreciate the value of others. In essence, awe fosters humility, and thus serves as one antidote to the human-centered hubris that has mired our thinking for centuries.

Human exceptionalism reaches back to classical antiquity and the emergence of the Judeo-Christian religious tradition. It has been formally codified and amplified throughout Western culture by other movements—including the Enlightenment, imperial colonialism, capitalism, and the Industrial Revolution. But perhaps most troublingly, human exceptionalism has also permeated the sciences, and thus the way many of us come to understand and relate to Nature today. More than any other ideology, this pervasive force cuts right across the political spectrum, shaping the current global industrialized economy and leaving many problems in its wake. The anthropocentric worldview creates hierarchies between humans and "nature." But as we've seen, it also perpetuates hierarchies among humans—rich over poor, white people over people of color, men over women, and so on. This Great Chain of Being has been used to systematically exclude classes of humans and the rest of life from self-determination, agency, dignity, and so much more. Dismantling these hierarchies is perhaps the biggest challenge of our time.

As I write, the United States is experiencing deepening divisions and escalating tensions around an upcoming presidential election. Political instability and social fragmentation are not unique to this country, but indicative of broader trends of growing nationalism. I might argue that a lot of the breakdown of the West right now parallels the story of human exceptionalism. American exceptionalism holds that the United States has a unique history and mission to transform the world, giving it superiority over other nations. But the rallying cry "America first" is just as hubristic as the rallying cry "Humans first." Both ideologies require us to move beyond the hierarchical paradigms that have too long constrained our manner of thinking.

Yet unlike national exceptionalism, beliefs in human exceptionalism are often *invisible*. Not necessarily because they are hidden or impossible to find, but because they are rarely explicitly taught, articulated, or scrutinized. They function as unspoken assumptions that dictate the behavior of individuals, corporations, nation-states, and other cultural institutions. And it is from this invisibility—this "matter-of-fact" sense that we are "obviously" superior to other species—that human exceptionalism draws its power. As Derrick

Jensen aptly puts it, "Unquestioned beliefs are the real authorities of any culture."

Unchecked beliefs in human exceptionalism take many forms but fundamentally assume that humans possess uniquely complex ways of being, thinking, and feeling—rendering other species' capacities, and thus their lives, inferior. But as we've seen, research often stacks the deck against these other species or measures their abilities according to human standards, biasing comparisons from the start. And despite this compulsion to differentiate, there is so much we share with other forms of life, many of whom even outshine us in certain feats. If intelligence is defined as solving novel problems in one's environment, then other species certainly possess it (and for all our vaunted intellect, we still struggle to resolve many environmental problems like global warming). They usefully expand our conventional definitions of intelligence and success, prompting us to reconsider the fundamental essence of mind, evolution, and who we are. Why should acknowledging the many contributions of other species diminish our own? Yes, humans arguably specialize in many feats. But doesn't every form of life? Aren't all species special then? And wise in their own right? "The anthropocentric worldview doesn't hold up very well to questions," quipped a student in his final paper for my class.

From charged debates over whether fish feel pain to whether plants are intelligent and even conscious; from heated disagreements over whether chimpanzees have a full-fledged theory of mind to whether songbirds have true language—scientists have spilled a lot of ink over such questions, and yet it sometimes seems that we are no closer to a consensus.

These disputes typically get framed as disagreements over semantics and scientific methods. But to me they have increasingly seemed like clashes over worldviews, over the narratives we construct and the discourses we subscribe to, often subconsciously. I now believe that some of the biggest scientific controversies today are a battle over assumptions regarding human exceptionalism, and a willingness or unwillingness to abandon those assumptions in the face of findings that undermine them.

Fortunately, history is filled with researchers who dared to make that leap. As we've seen, dissidents who challenge the anthropocentric paradigm are often initially not taken seriously. They are sometimes seen as misguided and eccentric; their evidence considered dubious and far-fetched. In my field, Jane Goodall's experiences also spring to mind. Today we take it for granted that chimpanzees have personalities and complex behavioral repertoires including tool use. But at one point, her observations of these qualities were deemed unscientific, anthropomorphic, and even sentimental. Goodall famously had papers rejected from journals for using chimpanzee names (rather than numeric codes), gender pronouns (instead of the conventional "it"), and words like "culture" to describe their communities. In a society dominated by human exceptionalism, thinking otherwise can be incredibly inconvenient. But history shows that, while initially disputed, discoveries challenging the ideology of human supremacy often later became widely accepted.

Every generation inherits a worldview it struggles to see past. The Copernican revolution revealed that humans were not the center of the cosmos. The Darwinian revolution showed that humans were one species among many, evolved from common origins. We are amid another revolution in how we understand ourselves in relation to the rest of Nature—one challenging the remaining strongholds of anthropocentrism in Western science. Just as previous revolutions have enriched our worldview and opened new and exciting avenues for scientific discovery, this revolution is transforming the science and the stories we tell about our place in the world.

Humans love stories. Storytelling is one of the oldest and most universal endeavors of our species, which helps explain our modern addiction to Netflix, podcasts, and social media. But as cultural historian Thomas Berry suggested, the anthropocentric story is no longer working for us. In "The New Story," Berry envisioned a cosmology in which humans are integral parts of the universe, not as sovereigns but as caring partners of the more-than-human world. As we've seen, this cosmology is far better supported by the evidence—our interdependence is evident in everything from the air we breathe to the cells that constitute our bodies. We've also learned that this "new" story need not be built from scratch—it is deeply rooted in Indigenous and other cultural traditions (even within the Western canon, though often hidden from view).

It is tempting to conclude this book with a single recommendation on how to live. I've often been asked if I have a hidden formula or a step-by-step list for moving toward a less anthropocentric world. But what I hope I've convinced you of is that we cannot continue to overlook the story of human exceptionalism. Of course, there is much more work to do to address the ecological crisis, but unlearning this deeply ingrained story is among the most important tasks at hand.

If you had one hour to save the world, how would you spend that hour? Albert Einstein reputedly answered that he would spend fifty-five minutes defining the problem and then five minutes solving it. We have spent a great deal of time identifying (though not always deploying) potential solutions—shifting to renewable energy, halting deforestation, restoring biodiversity, slowing population growth. But have we truly understood the underlying problem? If we take Einstein's insight to heart, grappling with the ecological crisis means dismantling not only systems of exploitation and destruction but also the very worldview that makes them possible. Western society has spent a long time convincing us that this worldview is a fact, and thus "solutions" too often operate through it. This is precisely why we need another story. As Berry articulated, "Even though this paradigm is no longer effective in dealing with the most basic issues of the present, there is a tendency to continue problem-solving within the paradigm rather than an effort to change the paradigm as the only way of dealing with the problems."

These are very hard problems, but many things can be done to move us closer to a less anthropocentric paradigm, and each of us can find some role to play, a task to undertake, that will shoulder a part of this immense collective reimagining. And while it might sound unrealistic or even utopian to some, this process of reworlding is already taking place in diverse parts of the globe, especially outside Western social constructs, and even within them.

It's happening in wildlife corridors and habitat restoration projects—from efforts to restore the Red Sea's native coral reefs to the miraculous comeback of Rwanda's mountain gorillas; from the revival of the American chestnut to pollinator-recovery projects in Brazil. It's underway in initiatives like the Mother Tree Project in British Columbia, where ecologists are collaborating with First Nations people to translate what we know about the sociality and intelligence of plants into forest renewal practices that protect biodiversity and promote carbon storage. It's unfolding in courtrooms, where lawyers are challenging anthropocentric legal structures by fighting for the rights of other species and ecosystems, while grassroots movements like Extinction Rebellion demand urgent climate action from governments. In the Netherlands, where I currently write, the Partij voor de Dieren (Party for the Animals) is the fastest-growing political party in the country, advocating strategically for animal interests and environmental protections. Meanwhile, degrowth and circular economic frameworks are cutting waste and pollution by promoting recycling, repair, and local production—reducing ecological footprints and enhancing community resilience. Reworlding is the specialty of artists, filmmakers, and writers who are decentering human perspectives and exploring alternative human-nature relationships through creative expression. Design practices are now embracing multispecies collaboration, as seen in projects like *I.N.S.E.C.T. Wall Twin*, an architectural installation that supports coexistence between local insects, fungi, and humans. Coexistence efforts are thriving in nature-based schools and ecovillages; in religious spaces and communities seeking to reconnect with the land. They're taking hold in herbalism programs, community gardens, and regenerative agriculture initiatives, where less anthropocentric practices are being cultivated every day. Consider "do-nothing farming" (a.k.a. "the Fukuoka Method," named for the Japanese farmer and philosopher who developed it), which recognizes a natural intelligence at work in the land, emphasizing minimal human intervention and allowing the ecosystem to balance itself and improve soil health naturally.

In my classroom, we're crafting this new story together. Since human exceptionalism isn't typically taught directly, unlearning it is more effective through conversation than instruction, which is why my course is largely discussion-based. With each group of students, another kind of microculture begins to emerge, where human-centered beliefs, norms, values, and language are reconsidered. Just as I discovered on my own path of unlearning human exceptionalism, my students begin to recognize how deeply this bias is embedded in our science. They, too, come to see that studies concluding that humans are exceptional in some capacity often establish little but the anthropocentric prejudices of the researchers conducting the work. And once you see this bias, it's impossible to unsee. For many of us, it becomes a revelation, another perspective on the world that breathes new life into our surroundings. Awe is no longer reserved for grand spectacles like a gorilla charge; it can be found in the reticulated patterns of a single leaf, the playful whirl of the wind, the hum of cicadas reverberating across the terrain. This shift in perception reveals grandeur in what might otherwise seem ordinary, making us wonder and feel connected to a broader, earthly narrative. The hunger for these experiences and conversations among students is palpable, and only mounting.

Among colleagues, I have also felt an unspoken bond building between those of us from various fields who regularly challenge anthropocentrism in our work. There are poets, writers, physicists, philosophers, theologians, artists, historians, decolonial scholars, anthropologists, ecologists, and many others. This growing and diverse movement goes by many names—biocentrism, posthumanism, new animism. In reality, it is less a formal movement than a shared vision among those who defy the commodification of the natural world and the tendency to mechanize life itself. Instead, they seek a more profound understanding of and connection with the earth and its myriad sentient inhabitants. They apprentice themselves to the land and local ecologies rather than attempting to master and control them. They suspect that we must eventually begin to scale down our technological ambitions, our blind commitment to the lure of progress. These are humans who are open to unlearning. Who are continually remembering. Who have, to borrow Robinson Jeffers's phrase, "fallen in love outward" with the world around them.

What might the world look like beyond human exceptionalism? Outdated anthropocentric narratives prevent people from imagining what else the world (and we ourselves) could be. And without a fresh vision or story to aspire to, many find themselves becoming disengaged and apathetic. Our experiences and beliefs about the world dictate how we behave in it. In other words, "worldview" isn't just something in our minds; it becomes a lived practice. Unlearning human exceptionalism is crucial because it transforms our relationship to the rest of Nature in ways that are embodied in our actions—a set of seemingly small acts that, on the cumulative scale of humanity, moves the world.

Despite years of researching and teaching these topics, I still sometimes catch myself falling into the trap of human exceptional-ism, and no doubt there are traces of this bias in these very pages. I am sure to regret these lapses as I continue to unlearn this stubborn ideology! I began writing this book in 2020, and by the time I fin-ished, I realized there is so much more I want to say and perhaps much I might express differently. I am still learning and unlearning what I believe to be true about the world, and how I want to partici-pate in it.

<hr />

Philosophers Valerie Tiberius and John Walker have examined why arrogance is considered a vice. They argue that arrogant people not only mistreat others but also impede their own pursuit of a good life, cit-ing two primary reasons. First, arrogance poses an obstacle to estab-lishing true friendships with others because it prevents the creation of relationships based on mutual enrichment. Second, these recipro-cal relationships serve as an invaluable source of self-knowledge—a "mirror onto the soul"—helping to reveal our true natures. While their work focuses on human relationships, the same reasoning can be easily applied to our interactions with other forms of life. Doing so unveils what is truly at stake in our estrangement from the rest of Nature: beyond limiting our understanding of these remarkable beings, human exceptionalism constrains a richer understanding of ourselves.

So what does it mean to be human? Our first hint might come from the word "human" itself—which derives from the root word *humus*, meaning "earth." To be human thus means to be of the earth, not apart from or better than any of the other beings with whom we share this planet. So why do we keep insisting otherwise? Is there an alternative approach to this age-old defense mechanism?

According to cultural ecologist David Abram, the answer may lie in another word that shares the same ancestry: "humility," the state of being close to the earth. As Abram points out, such shared etymologies suggest that we are most human when we maintain a humble relationship with the world around us.

At a dinner party years ago, I was asked which emotion or character trait I would choose to enhance in every person worldwide if I could administer a pill for it. Given my research on empathy, the group expected that to be my choice. I paused, because that didn't feel quite right. My friend proposed "awe," and I found myself agreeing. But on further reflection, today I might choose humility, a virtue that awe often inspires.

Humility involves holding a realistic, secure view of the self alongside an appreciation of the value and contribution of others. It is considered a foundational virtue because it counters selfish and socially disruptive inclinations like arrogance. Yet humility is one of the most understudied virtues in the social and psychological sciences. The handful of studies on it that exist indicate that humility brings various interpersonal benefits (such as healthier social relationships and increased altruism) and personal benefits (such as greater well-being and resilience). But again, in virtually every case, humility research has focused on how humans understand themselves vis-à-vis other humans. What if we applied this unsung virtue to how we relate to the rest of the living world, and to the science that seeks to understand it?

Let us now return to our titular arrogant ape. In classical Greek theater, humility stood in opposition to hubris, an excessive pride leading people to overestimate their own abilities. Hubris often ended in tragedy, as those who violated the natural order faced the wrath of Nemesis, the goddess who restored cosmic balance and justice. Our current crises—climate change, pandemics, and environmental degradation—can be viewed as Nature's way of imposing consequences for our hubris. But instead of trying to "solve" these crises with the same hubristic mindset that created them, we might benefit from embracing humility. The moment is thus ripe for anagnorisis—from the Greek for "recognition"—the profound realization that produces a change from ignorance to knowledge. Anagnorisis revealed fundamental truths about a character's identity or the nature of their situation and relationships, marking a critical turning point that brought about either the tragic downfall or transformation of the protagonist. Through the lens of this metaphor, this pivotal ecological moment can be seen either optimistically or pessimistically, but I favor neither. Instead, I tend toward hope. Humility means recognizing that we cannot understand everything and do not know what lies ahead, rather than clinging to false assurances or dire outlooks. Optimism and pessimism are probabilistic; they proclaim to know the odds, and await a better or worse future. Hope, on the other hand, centers on potential and uncertainty—it's about not knowing. In other words, hope is more aligned with humility.

Hope arises when we realize that human exceptionalism is not an inherent trait, not a bias we're born with. Rather, it's a role we've assumed thanks to a cultural story we've inherited, like characters in a play. The script is written, the stage is set, and we have cast ourselves as the protagonist (*Homo sapiens sapiens*, we've called ourselves, the wisest of the wise). Craving the spotlight, we aim to direct and control this performance instead of letting the true multispecies story unfold. Humility means accepting that Nature will ultimately play her part regardless of our plans. It's about embracing the reality that other species have vital roles to fill and that learning from them is the path to real wisdom. Humility of this sort may not come easily; it's something we must practice.

In its ancient Greek context, drama was a way of learning what it means to be human. But tragedy was never a story just about humans. The roots of tragedy involved masked figures—hybrids of animals and gods—who honored earth's cycles and rhythms in archaic festivals. These festivals explored the potential for human transformation and the connection between people and Nature. By participating in these rites, people transcended their daily lives, gained wisdom from the more-than-human world, and reconsidered their place in the cosmos.

The "arrogant ape" is merely one mask, a role we can choose to abandon. By shedding the mask, we open the way for a new story— one that is beckoning us, ready to be told.

Acknowledgments

WITHOUT THE SUPPORT, PATIENCE, AND GUIDANCE OF many people—human and otherwise—this book would not have been possible.

I am so grateful to my literary agent, Michelle Tessler, for her steady counsel. Thank you for believing in this book from the start and for getting it into the right hands. My sincere thanks go to my editors at Avery, Tracy Behar and Caroline Sutton, whose sharp editorial insights and enthusiasm have carried this book through. I extend further gratitude to everyone at Penguin Random House who helped bring this project to fruition, including Elisabeth Koyfman, Hannah Steigmeyer, Jess Morphew, Jamie Lescht, Maya Ono, Angie Boutin, Sally Knapp, and M. P. Klier. Many thanks to Bill Mayer for his exquisite cover art. I also wish to thank my editor at Little, Brown in the UK, Anna Kelly, for her early encouragement and feedback.

I am indebted to my mentor and collaborator Frans de Waal, who not only laid the groundwork for many of these concepts but also took a chance on me as a young scholar, time and again. And to my *karass*—Becca Franks, Monica Gagliano, and Barbara Smuts—thank you for giving me permission to explore these ideas, both as a human and as a scientist. So many of our collaborative discussions are folded into these pages. Special thanks to Jennifer Jacquet for nudging me to write this book (if it sparks angry e-mails, I'll know who to blame) and providing helpful feedback along the way. And to Tory Higgins for his dedication—both to his students and to a psychology beyond the human.

I am grateful to Harvard's Department of Human Evolutionary Biology for being my academic home while writing. Thank you to my colleagues there—especially Lys Alcayna-Stevens, Joyce Benenson, Julia Espinosa, Carole Hooven, Julie Lawrence, Dan Lieberman, Barbara Natterson-Horowitz, Isaac Schamberg, Martin Surbeck, Rachna Reddy, Liran Samuni, Emily Walco, Erin Wessling, and Richard Wrangham—for advice and stimulating conversations. This book has benefited from the work of many other research collaborators: Alice Baniel, Jake Brooker, Alecia Carter, Zanna Clay, Xuejing Du, Zoë Goldsborough, Elise Huchard, Kayla Kolff, Stephanie Kordon, Matthew Lewis, Teresa Romero, Maya Rossignac-Milon, Liesbeth Sterck, Edwin van Leeuwen, Debbie Walsh, and Peter Woodford, as well as numerous research assistants. Thank you for everything you do to bring more attention to animals' rich social and emotional lives.

Many thanks to the Tsaobis Baboon Project, Chimfunshi Wildlife Orphanage, and Lola ya Bonobo Sanctuary for providing beautiful field sites for my research, and for prioritizing the interests of the primates under their study and care. I am also grateful to the Templeton World Charity Foundation, the Carper Foundation, and the Fyssen Foundation for their generous support of my work.

Thank you to the Harvard Interspecies Dialogues Reading Group (formerly known as the Animism Reading Group!) and Plant Consciousness Reading Group for enriching discussions and community. A special thank-you to Natalia Schwien for welcoming me into these spaces, and for the gift of your friendship. And to other friends and colleagues via Harvard's Center for the Study of World Religions—David Abram, Nicole Mandell, Cadine Navarro, Rachael Petersen, Christina Seely, and Charlie Stang—for your invaluable support and wise words along the way.

To my wonderful students, who have brought these ideas to life in ways I could never have imagined possible—your wisdom is everywhere in this book. Special thanks to those from whom I have learned so much through conversations beyond the classroom, including Lucrecia Aguilar, Lena Ashooh, Esme Benson, Maddie Dowd, Aysha Emmerson, Campbell Erickson, Byron Hurlbut, Yash Kumbhat, Anissa Medina, Amelia Quezada, Rob Rosenthal, Sophia Scott, Jordan Villegas, and Morgan Whitten.

I have benefited from interactions with numerous other scholars while writing this book, including Colin Allen, Kristen Andrews, Mark Bekoff, Lucien Castaing-Taylor, Eileen Crist, Bathsheba Demuth, Roger Dube, Lori Gruen, Mathias Guenther, Catherine Marin, Sicelo Mbatha, Eva Meijer, Harriet Ritvo, and Harry Wels. And there are countless individuals—scientists, artists, poets, philosophers, and others—in the posthuman turn whose work I continue to admire from afar and who cannot all be listed here; instead, I wish to thank those whose works are cited in the references.

Further thanks to the tireless staff at Harvard Libraries, who provided essential support, especially during the COVID lockdowns. Their dedication—scanning and hauling books, and even showing genuine interest in my work—is much appreciated.

I'm thankful for the Massachusett people, on whose traditional, ancestral, and unceded territories Harvard is located and where I have resided while writing much of this book.

To all of the birds who cheered me on as I wrote, crafting entire homes for their families before I finished a chapter: thank you for keeping the writing process in perspective—and for reminding me of my debt to the world, and all of the beings who live in it. There are many beings with whom I've worked and lived over the years—Amber, Arly, Bear, Dudley, G.B., Ginny, Jake, Libby, Macduff, Munyinya, Nicki, Raimee, Venus, Willow, Winter, Zippy—who have imparted lessons that can't readily be put into words. I hope this book reciprocates in some small way.

Finally, thank you to my human relatives: Carry Zwitserlood and the de Breed family; Aunt Carol, for countless proofreads; Dad, who acts as my "unofficial" manager; and Mom, whose creativity helped this book from start to finish—your support means everything. To Lucas, for being on human exceptionalism patrol and for a love that brings out my inner child. And to Vera, for providing a hard deadline, and for changing my worldview anew.

Notes

CHAPTER ONE: THE HUMAN SUPERIORITY COMPLEX

1 **"What a piece of work":** William Shakespeare (1603/1877). *Hamlet,* act 2, scene 2. In Horace H. Furness (ed.), *A New Variorum Edition of Shakespeare, Vol. 1* (p. 159). J. B. Lippincott.

1 **Psychologists have shown:** Alfred Adler (1930). "Individual psychology." In Carl Murchison (ed.), *Psychologies of 1930* (pp. 395–405). Clark University Press.

2 **"Although we are animals":** Roger Scruton (March 6, 2017). "If we are not just animals, what are we?" *New York Times.* https://www.nytimes.com/2017/03/06/opinion/if-we-are-not-just-animals-what-are-we.html.

2 **"We cannot satisfactorily":** Adam Rutherford (September 21, 2018). "The human league: What separates us from other animals?" *The Guardian.* https://www.theguardian.com/books/2018/sep/21/human-instinct-why-we-are-unique.

2 **"It seems obvious":** Thomas Suddendorf (November 21, 2013). "Are we really different from other animals?" CNN. https://www.cnn.com/2013/11/21/health/animals-humans-gap/index.html.

2 **"why we can cooperate":** Guy Raz (March 4, 2016). "Why did humans become the most successful species on earth?" *Ted Radio Hour.* NPR. https://www.npr.org/transcripts/468882620?t=1600086855866.

2 **"Maybe the reason":** Melissa Healy (December 20, 2016). "Chimpanzees may be helpful, but humans are the only primates that are kind to others, study suggests." *Los Angeles Times.* https://www.latimes.com/science/sciencenow/la-sci-sn-kindness-chimpanzees-humans-20161220-story.html.

3   **opened his keynote address:** Michael Tomasello (May 23, 2019). *Becoming Human: A Theory of Ontogeny.* Association for Psychological Science Annual Convention. https://www.youtube.com/watch?v=BNbeleWvXyQ.

3   **dubbed "humaniqueness" by:** Marc Hauser (November 12, 2008). *The Seeds of Humanity.* Tanner Lectures on Human Values, delivered at Princeton University.

5   **"Man in his arrogance":** Charles Darwin (1837/1974). C Notebook. In Howard E. Gruber and Paul H. Barrett (eds.), *Darwin on Man* (pp. 196–197). Wildwood House.

7   **Western, educated, industrialized, rich, and democratic—WEIRD:** Joseph Henrich, Steven J. Heine, and Ara Norenzayan (2010). "The WEIRDest people in the world?" *Behavioral and Brain Sciences* 33: 61–83.

8   **"Study: Dolphins Not So Intelligent":** *The Onion* (February 15, 2006). https://www.theonion.com/study-dolphins-not-so-intelligent-on-land-1819 568299.

11   **More than 90 percent of the earth's soil:** Food and Agriculture Organization (May 15–17, 2019). Global Symposium on Soil Erosion, "Key messages." https://www.fao.org/about/meetings/soil-erosion-symposium/key-messages/en.

11   **Thirty percent of global forest cover:** World Resources Institute. Accessed March 30, 2021. https://www.f6s.com/worldresourcesinstitutewri/about.

11   **Earth's temperature has risen:** Rebecca Lindsey and Luann Dahlman (January 18, 2024). "Climate change: Global temperature." National Oceanic and Atmospheric Administration. https://www.climate.gov/news-features/understanding-climate/climate-change-global-temperature.

11   **Ocean acidification has been occurring:** European Environment Agency. "Ocean acidification." Accessed June 13, 2020. https://www.eea.europa.eu/data-and-maps/indicators/ocean-acidification-2/assessment.

11   **Wild animal populations:** World Wildlife Fund. *Living Planet Report 2020.* Accessed June 14, 2021. https://f.hubspotusercontent20.net/hubfs/4783129/LPR/PDFs/ENGLISH-FULL.pdf.

11   **Pollinators on whom:** Intergovernmental Science-Policy Platform on Biodiversity and Ecosystem Services (2016). *Assessment Report on Pollinators, Pollination and Food Production.* Accessed June 14, 2021. https://ipbes.net/assessment-reports/pollinators.

11   **"I could go on":** Paul Kingsnorth (2017). *Confessions of a Recovering Environmentalist and Other Essays* (p. 2). Graywolf Press.

12   **time period the Anthropocene:** Paul J. Crutzen (2002). "Geology of mankind." *Nature* 415 (6867): 23.

12   **criticism from various scholars:** For example, Andreas Malm and Alf Hornborg (2014). "The geology of mankind? A critique of the Anthropocene narrative." *Anthropocene Review* 1 (1): 62–69; Eileen Crist (2013). "On the poverty of our nomenclature." *Environmental Humanities* 3: 129–147.

CHAPTER TWO: THE UNLEARNING CURVE

16   **the term "biophilia":** Edward O. Wilson (1984). *Biophilia*. Harvard University Press. The term was coined by sociologist Erich Fromm: Erich Fromm (1964). *The Heart of Man: Its Genius for Good and Evil*. Harper and Row.

16   **In one of my favorite videos:** Weirdo Ultimaes (April 17, 2015). "Budding vegetarian makes mom cry with his love of animals." https://www.youtube.com/watch?v=oL8jz_Dl2Pc.

18   **In one 2021 study:** Matti Wilks, Lucius Caviola, Guy Kahane, and Paul Bloom (2021). "Children prioritize humans over animals less than adults do." *Psychological Science* 32 (1): 27–38.

19   **children rated farm animals:** Scott Plous (1993). "Psychological mechanisms in the human use of animals." *Journal of Social Issues* 49: 11–52.

20   **Species used in agriculture:** Peter Singer (2002). *Animal Liberation: A New Ethics for Our Treatment of Animals*. HarperCollins.

20   **People eat "chicken":** I borrow this suggestion from Plous (1993). "Psychological mechanisms in the human use of animals" (p. 17).

20   **Researchers in cognitive linguistics:** See William Croft and D. Alan Cruse (2004). *Cognitive Linguistics*. Cambridge University Press.

21   **Sociologist Eileen Crist calls:** Eileen Crist (2013). "On the poverty of our nomenclature." *Environmental Humanities* 3: 129–147.

21   **Even dictionaries, which are seen:** See Reinhard Heuberger (2003). "Anthropocentrism in monolingual English dictionaries: An ecolinguistic approach to the lexicographic treatment of faunal terminology." *AAA: Arbeiten aus Anglistik und Amerikanistik* 28 (1): 93–105.

21   **ask Google to define:** Unfortunately, we are also imparting these anthropocentric prejudices to mainstream AI applications. For example, WordNet and other annotation structures for popular image datasets contain speciesist terms like "hog," "porker," and "livestock." See Thilo Hagendorff, Leonie N. Bossert, Yip Fai Tse, and Peter Singer (2023). "Speciesist bias in AI: How AI applications perpetuate discrimination and unfair outcomes against animals." *AI and Ethics* 3: 717–734. See also Rachel Teng (November 3, 2023). "The AI bias that's often overlooked: Speciesism." Sentient Media. https://sentientmedia.org/ai-bias-speciesism.

21   **Wesleyan psychologist Scott Plous:** Scott Plous (2003). "Is there such a thing as prejudice towards animals?" In Scott Plous (ed.), *Understanding Prejudice and Discrimination* (pp. 509–528). McGraw-Hill Higher Education.

22   **advised authors to substitute words:** Described in Susan E. Lederer (1992). "Political animals: The shaping of biomedical research literature in twentieth-century America." *Isis* 83: 61–79.

22   **"When gender is known":** Jane Goodall et al. (2021). "Joint open letter to the Associated Press calling for a change in animal pronouns." Accessed September

24, 2024. https://www.idausa.org/assets/files/assets/uploads/pdf/openletterap
stylebook.pdf.

23   **Jacques Derrida once elaborated:** Jacques Derrida (2008). *The Animal That Therefore I Am.* Fordham University Press.

23   **"The arrogance of English":** Robin Wall Kimmerer (2012). "Learning the grammar of animacy" (p. 9). *The Leopold Outlook* (Winter): 1–9.

25   **Interestingly, research indicates:** Luke McGuire, Sally B. Palmer, and Nadira S. Faber (2023). "The development of speciesism: Age-related differences in the moral view of animals." *Social Psychological and Personality Science* 14 (2): 228–237.

25   **Some studies suggest:** Aurélien Miralles, Michel Raymond, and Guillaume Lecointre (2019). "Empathy and compassion toward other species decrease with evolutionary divergence time." *Scientific Reports* 9 (1): 1–8.

27   **Dorian Solot and Arnold Arluke:** Dorian Solot and Arnold Arluke (1997). "Learning the scientist's role: Animal dissection in middle school." *Journal of Contemporary Ethnography* 26: 28–54.

27   **"One of the skills":** Quoted in Nancy Averett (January 27, 2020). "High school dissections are a science class tradition, but are they doing more harm than good?" *Discover.* https://www.discovermagazine.com/the-sciences/the-ar gument-against-high-school-animal-dissections.

28   **Kopnina's research showed:** Helen Kopnina, Michael Sitka-Sage, Sean Blen-kinsop, and Laura Piersol (2018). "Moving beyond innocence: Educating children in a post-nature world." In Amy Cutter-Mackenzie-Knowles, Karen Malone, and Elisabeth Barratt Hacking (eds.), *Research Handbook on Childhoodnature* (pp. 603–621). Springer.

29   **"History is the story":** John M. Roberts (2014). *The Penguin History of the World* (6th ed., p. 1). Penguin Books.

30   **the "species" category:** James Mallet, Fernando Seixas, and Yuttapong Tha-wornwattana (2022). "Species, concepts of." In Samuel M. Scheiner (ed.), *Encyclopedia of Biodiversity* (3rd ed., pp. 531–545). Academic Press.

30   **creatures like earthworms:** For a wonderful exposition of this, see Eileen Crist (2002). "The inner life of earthworms: Darwin's argument and its implications." In Marc Bekoff, Colin Allen, and Gordon M. Burghardt (eds.), *The Cognitive Animal: Empirical and Theoretical Perspectives on Animal Cognition* (pp. 3–8). MIT Press.

31   **"Nothing in biology":** Theodosius Dobzhansky (1973). "Nothing in biology makes sense except in the light of evolution." *American Biology Teacher* 35 (3): 125–129.

31   **"all species are unique":** Theodosius Dobzhansky (1955). *Evolution, Genetics, and Man* (p. 12). John Wiley.

31   **Stephen Jay Gould, who taught:** See, for instance, Stephen Jay Gould (1989). *Wonderful Life: The Burgess Shale and the Nature of History.* Norton.

31   **presented as progressive:** Sean Nee (2005). "The great chain of being." *Nature* 435: 429.

31   **"it is absurd to talk of":** Charles Darwin (1837). Notebook B: (Transmutation, 1837–1838). *Darwin Online*, CUL-DAR121. Accessed April 21, 2022. http://darwin-online.org.uk/content/frameset?itemID=CUL-DAR121 .-&viewtype=side&pageseq=1.

32   **Are frogs more closely:** I thank the following paper for this suggestion, complemented by a set of illuminating illustrations: David A. Baum, Stacey Dewitt Smith, and Samuel S. S. Donovan (2005). "The tree-thinking challenge." *Science* 310: 979–981.

32   **"The vast majority of modern fish":** Becca Franks (December 22, 2020). "Fish in the 21st century: The good, the bad, and the hopeful." *Nautilus*. https://nautil.us/fish-in-the-21st-century-the-good-the-bad-and-the-hopeful -11804. See also Becca Franks, Christine Webb, Monica Gagliano, and Barbara Smuts (2022). "Looking up to animals and other beings: What the fishes taught us." In Melanie Challenger (ed.), *Animal Dignity: Reflections on Our Respect for Other Species* (pp. 229–238). Bloomsbury Academic.

33   **"When we come to realize":** Robert J. O'Hara (1992). "Telling the tree: Narrative representation and the study of evolutionary history" (p. 157). *Biology and Philosophy* 7: 135–160.

33   **Carey's influential model:** Susan Carey (1985). *Conceptual Change in Childhood*. MIT Press.

34   **Carey found that young:** Carey (1985). *Conceptual Change in Childhood* (pp. 126–135).

34   **also known as WEIRD:** Joseph Henrich, Steven J. Heine, and Ara Norenzayan (2010). "The WEIRDest people in the world?" *Behavioral and Brain Sciences* 33: 61–83.

34   **anthropologists performed:** Reviewed in Douglas Medin and Scott Atran (2004). "The native mind: Biological categorization and reasoning in development and across cultures." *Psychological Review* 111 (4): 960–983.

35   **one thousand corporate logos:** See Colin Marshall (June 17, 2022). "The cracked wisdom of Dril." *New Yorker*. https://www.newyorker.com/culture /rabbit-holes/the-cracked-wisdom-of-dril.

35   **biological categories for WEIRD adults:** Described in Henrich, Heine, and Norenzayan (2010). "The WEIRDest people in the world?" (p. 67).

35   **One study compared:** Kayoko Inagaki (1990). "The effects of raising animals on children's biological knowledge." *British Journal of Developmental Psychology* 8 (2): 119–129.

36   **To test this, Patricia Herrmann:** Patricia Herrmann, Sandra R. Waxman, and Douglas L. Medin (2010). "Anthropocentrism is not the first step in children's reasoning about the natural world." *Proceedings of the National Academy of Sciences of the United States of America* 107 (22): 9979–9984.

36 **One case study:** Scott Atran et al. (2002). "Folkecology, cultural epidemiology, and the spirit of the commons." *Current Anthropology* 43 (3): 421–450.

37 **"nature-deficit disorder":** Richard Louv (2008). *Last Child in the Woods: Saving Our Children from Nature-Deficit Disorder.* Algonquin.

37 **"shifting baseline syndrome":** Daniel Pauly (1995). "Anecdotes and the shifting baseline syndrome of fisheries." *Trends in Ecology and Evolution* 10: 430.

37 **"environmental generational amnesia":** Peter H. Kahn Jr. (2002). "Children's affiliations with nature: Structure, development, and the problem of environmental generational amnesia." In Peter H. Kahn Jr. and Stephen R. Kellert (eds.), *Children and Nature* (pp. 93–116). MIT Press.

38 **One telling example:** Samuel T. Turvey et al. (2010). "Rapidly shifting baselines in Yangtze fishing communities and local memory of extinct species." *Conservation Biology* 24 (3): 778–787.

39 **Whereas children watched:** Frans de Waal, e-mail to author, November 14, 2021.

39 **"Children in the industrialized world":** John Berger (1980). "Why look at animals?" In *About Looking* (p. 22). Pantheon Books.

40 **"It is common to hear":** Jenny Kendler. "Tell it to the birds." Accessed December 2, 2024. https://jennykendler.com/section/402442-Tell%20it%20to%20the%20Birds.html.

41 **So we turn away:** That is, if we can—studies have shown that slaughterhouse workers who can't turn away suffer psychological trauma and are themselves also victims of this exploitative system. Jessica Slade and Emma Alleyne (2021). "The psychological impact of slaughterhouse employment: A systematic literature review." *Trauma, Violence, & Abuse* July: 1–12.

CHAPTER THREE: A COMPULSION TO CONTRAST

43 **called this the "Differential Imperative":** John Rodman (1980). "Paradigm change in political science: An ecological perspective." *American Behavioral Scientist* 24 (1): 49–78.

44 **early human societies:** For a provocative reinterpretation of human history in this regard, see David Graeber and David Wengrow (2021). *The Dawn of Everything: A New History of Humanity.* Farrar, Straus and Giroux.

44 **Eileen Crist points out:** Eileen Crist (2019). *Abundant Earth: Toward an Ecological Civilization* (pp. 52–53). University of Chicago Press.

44 **known as the *scala naturae*:** See J. David Archibald (2014). *Aristotle's Ladder, Darwin's Tree: The Evolution of Visual Metaphors for Biological Order.* Columbia University Press.

45 **As scholars like Gary Steiner:** Gary Steiner (2005). *Anthropocentrism and Its Discontents: The Moral Status of Animals in the History of Western Philosophy.* University of Pittsburg Press.

45 **"After the birth":** Aristotle (ca. 350 B.C./2014). Politics. In Jonathan Barnes (ed.), *The Complete Works of Aristotle, Vol. 2: The Revised Oxford Translation* (pp. 1993–1994). Princeton University Press.

46 **progressionist linear language:** Emanuele Rigato and Alessandro Minelli (2013). "The Great Chain of Being is still here." *Evolution: Education and Outreach* 6: 1–6.

46 **idea of the Great Chain of Being:** Arthur O. Lovejoy (1976). *The Great Chain of Being: A Study of the History of an Idea.* Harvard University Press.

46 **Anticipating the biblical story:** I borrow this suggestion from Crist (2019). *Abundant Earth* (p. 53).

46 **"Many the wonders":** Quoted in Hans Jonas (1974). *Philosophical Essays* (p. 122). Prentice-Hall.

47 **"And God said":** The Bible (English Standard Version). Genesis 1:28.

47 **"the orthodox Christian arrogance":** Lynn White Jr. (1967). "The historical roots of our ecological crisis" (p. 1207). *Science* 55: 1203–1207.

48 **"By gradual stages":** White Jr. (1967). "The historical roots of our ecological crisis" (p. 1205).

50 **"victory of Christianity over paganism":** White Jr. (1967). "The historical roots of our ecological crisis" (p. 1205).

50 **"the great civilising influence":** Quoted in Keith Thomas (1983). *Man and the Natural World: Changing Attitudes in England 1500–1800* (p. 23). Penguin.

50 **alternative interpretations of biblical:** See Matthew Scully (2002). *Dominion: The Power of Man, the Suffering of Animals, and the Call to Mercy.* St. Martin's Griffin; Mark I. Wallace (2018). *When God Was a Bird.* Fordham University Press.

50 ***Catechism of the Catholic Church:*** *Catechism of the Catholic Church for the United States of America* (1997). CatholicCulture.org. Accessed September 14, 2024. https://www.catholicculture.org/culture/library/catechism/index.cfm?recnum=6253.

51 **International Committee of Human Dignity:** "Universal Declaration of Human Dignity." Accessed September 14, 2024. http://www.dignitatishuma nae.com/wp-content/uploads/2015/08/Declaration.pdf.

51 **2019 Gallup poll:** Gallup. "Evolution, Creationism, Intelligent Design." Accessed September 14, 2024. https://news.gallup.com/poll/21814/evolution -creationism-intelligent-design.aspx.

51 **Buddhists do not necessarily regard:** Alex Bruce (2018). "Buddhism: Paradox and practice—morally relevant distinctions in the Buddhist characterization of animals." In Andrew Linzey and Clair Linzey (eds.), *The Routledge Handbook of Religion and Animal Ethics* (pp. 43–55). Routledge.

53 **"no experience of any kind":** Quoted in Steiner (2005). *Anthropocentrism and Its Discontents* (p. 147).

54 **"I perceived it to be":** René Descartes (1637/1927). *Discourse on Method* (p. 66). Open Court Publishing.

54 **"The mechanical inventions":** Francis Bacon (1607/1964). "Thoughts and conclusions on the interpretation of nature or a science of productive works." In Benjamin Farrington, *The Philosophy of Francis Bacon* (p. 93). Liverpool University Press.

55 **"a shore fit for Pandemonium":** Quoted in Jonathan Weiner (1994). *The Beak of the Finch* (p. 13). Knopf.

55 **He once wrote that:** Charles Darwin (1845). *Journal of Researches into the Natural History and Geology of the Countries Visited During the Voyage of H.M.S. Beagle Round the World, Under the Command of Capt. Fitz Roy, R.N.* (2nd ed., p. 398). John Murray.

55 **Another time, he pulled:** Darwin (1845). *Journal of Researches (*p. 388).

55 **one-quarter of his notes:** GoGalapagos. "Charles Darwin and the Galapagos Islands." Accessed December 19, 2020. https://www.gogalapagos.com/charles-darwin-galapagos.

56 **"mystery of mysteries":** Charles Darwin (1859/1902). *On the Origin of Species* (p. 25). American Home Library.

56 **"like confessing a murder":** Darwin Correspondence Project. "Letter no. 729." Accessed April 4, 2022. https://www.darwinproject.ac.uk/letter/?docId=letters/DCP-LETT-729.xml.

57 **"The main conclusion":** Charles Darwin (1871/1981). *The Descent of Man, Vol. 2* (p. 404). Princeton University Press.

57 **"Ascent of Man" measure:** Nour Kteily, Emile Bruneau, Adam Waytz, and Sarah Cotterill (2015). "The ascent of man: Theoretical and empirical evidence for blatant dehumanization." *Journal of Personality and Social Psychology* 109 (5): 901–931.

57 **"This generates a curious paradox":** Melanie Challenger (April 6, 2021). "The joy of being animal." *Aeon.* https://aeon.co/essays/to-be-fully-human-we-must-also-be-fully-embodied-animal.

58 **racist and sexist view:** See Agustín Fuentes (2021). "'The Descent of Man,' 150 years on." *Science* 372 (6544): 769 and associated commentary.

58 **"Despite Darwin," wrote Lynn White Jr.:** White Jr. (1967). "The historical roots of our ecological crisis" (p. 1206).

58 **"Now we must redefine":** Quoted at Jane Goodall Institute. "Our legacy of science." Accessed April 30, 2022. https://janegoodall.org/our-story/our-legacy-of-science.

58 **"the most satisfactory":** Kenneth Oakley (1956). "The earliest tool-makers." *Antiquity* 30 (117): 4–8.

59 **our avian cousins:** For an exhaustive review, see Robert W. Shumaker, Kristina R. Walkup, and Benjamin B. Beck (2011). *Animal Tool Behavior* (pp. 35–58). John Hopkins University Press.

60 **researchers presented bird participants:** Alex A. S. Weir, Jackie Chappell,

and Alex Kacelnik (2002). "Shaping of hooks in New Caledonian crows." *Science* 297 (5583): 981.

60　**Later work showed:** Barbara C. Klump, Shoko Sugasawa, James J. H. St. Clair, and Christian Rutz (2015). "Hook tool manufacture in New Caledonian crows: Behavioural variation and the influence of raw materials." *BMC Biology* 13 (1): 1–15.

60　**Octopuses gather coconut shells:** Julian K. Finn, Tom Tregenza, and Mark D. Norman (2009). "Defensive tool use in a coconut-carrying octopus." *Current Biology* 19 (23): 1069–1070.

60　**Elephants pick up:** Suzanne Chevalier-Skolnikoff and Jo Liska (1993). "Tool use by wild and captive elephants." *Animal Behaviour* 46: 209–219.

61　**tree crickets construct:** Natasha Mhatre, Robert Malkin, Rittik Deb, Rohini Balakrishnan, and Daniel Robert (2017). "Tree crickets optimize the acoustics of baffles to exaggerate their mate-attraction signal." *eLife* 6: e32763.

61　**"that complex whole":** Edward B. Tylor (1871). *Primitive Culture* (p. 1). John Murray.

61　**research team of Kinji Imanishi:** See Tetsuro Matsuzawa and William C. McGrew (2008). "Kinji Imanishi and 60 years of Japanese primatology." *Current Biology* 18 (14): R587–R591.

62　**conservation initiatives underway:** S. J. Ryan (2006). "The role of culture in conservation planning for small or endangered populations." *Conservation Biology* 20 (4): 1321–1324.

62　**"A desire to take medicine":** William Osler (1891). "Recent advances in medicine" (p. 170). *Science* 17: 170–171.

62　**surge of research:** Reviewed in Joel Shurkin (2014). "Animals that self-medicate." *Proceedings of the National Academy of Sciences of the United States of America* 1112 (49): 17339–17341.

62　**One 2022 report:** Alessandra Mascaro, Lara M. Southern, Tobias Deschner, and Simone Pika (2022). "Application of insects to wounds of self and others by chimpanzees in the wild." *Current Biology* 32 (3): R112–R113.

63　**"an adult male":** Quoted in Ashley Strickland (February 7, 2022). "Chimpanzees apply 'medicine' to each others' wounds in a possible show of empathy." CNN. https://www.cnn.com/2022/02/07/world/chimpanzee-insects-wounds-scn/index.html.

64　**chimps try to protect:** Christine E. Webb, Kayla Kolff, Xuejing Du, and Frans de Waal (2020). "Jealous behavior in chimpanzees elicited by social intruders." *Affective Science* 1 (4): 199–207.

64　**"Why should our nastiness":** Stephen Jay Gould (1980). *Ever Since Darwin* (p. 261). Penguin.

65　**other species appear more rational:** See, for instance, Victoria Horner and Andrew Whiten (2005). "Causal knowledge and imitation/emulation switching in chimpanzees (*Pan troglodytes*) and children (*Homo sapiens*)." *Animal Cogni-*

tion 8 (3): 164–181; Angie M. Johnston, Paul C. Holden, and Laurie R. Santos (2017). "Exploring the evolutionary origins of overimitation: A comparison across domesticated and non-domesticated canids." *Developmental Science* 20: e12460.

65 **chimpanzees maximize their benefits:** Keith Jensen, Josep Call, and Michael Tomasello (2007). "Chimpanzees are rational maximizers in an ultimatum game." *Science* 318 (5847): 107–109.

65 **animals can be just as irrational:** Thomas R. Zentall (2016). "When humans and other animals behave irrationally." *Comparative Cognition and Behavior Reviews* 11 (1): 25–48.

65 **"Language is the last":** Ferris Jabr (May 12, 2017). "Can prairie dogs talk?" *New York Times.* https://www.nytimes.com/2017/05/12/magazine/can-prairie -dogs-talk.html.

66 **Vervet monkeys gained:** Robert M. Seyfarth, Dorothy L. Cheney, and Peter Marler (1980). "Vervet monkey alarm calls: Semantic communication in a free-ranging primate." *Animal Behaviour* 28 (4): 1070–1094.

66 **prairie dogs specify:** Con Slobodchikoff (2012). *Chasing Doctor Dolittle: Learning the Language of Animals* (pp. 54–62). St. Martin's Press.

66 **Japanese great tits:** Toshitaka N. Suzuki, David Wheatcroft, and Michael Griesser (2016). "Experimental evidence for compositional syntax in bird calls." *Nature Communications* 7 (10986).

66 **Starlings, it turns out:** Timothy Q. Gentner, Kimberly M. Fenn, Daniel Margoliash, and Howard C. Nusbaum (2006). "Recursive syntactic pattern learning by songbirds." *Nature Communications* 440 (7088): 1204–1207.

67 **introduces something called "The Sentence":** Daniel Gilbert (2006). *Stumbling on Happiness* (pp. 3–4). Knopf.

67 **"If hypotheses about human uniqueness":** Colin A. Chapman and Michael A. Huffman (2018). "Why do we want to think humans are different?" (p. 4). *Animal Sentience* 163: 1–8.

68 **"a sort of natural deficiency":** Aristotle (ca. 350 B.C./2014). Generation of Animals. In Jonathan Barnes (ed.), *The Complete Works of Aristotle, Vol. 1: The Revised Oxford Translation* (p. 1199). Princeton University Press.

68 **"Beasts in the skin of man":** Quoted in Thomas (1983). *Man and the Natural World* (p. 42).

69 **"In this primitive":** Quoted in Thomas (1983). *Man and the Natural World* (p. 42).

69 **"possessed and wrongfully":** Quoted in Thomas (1983). *Man and the Natural World* (p. 42).

69 **"the greatest part":** Quoted in Thomas (1983). *Man and the Natural World* (p. 42).

69 **"they act like wolves":** Quoted in Thomas (1983). *Man and the Natural World* (p. 47).

69 **"are but brutes":** Quoted in Thomas (1983). *Man and the Natural World* (pp. 43–44).

70 **characterized as lacking:** Reviewed in Nick Haslam (2006). "Dehumanization: An integrative review." *Personality and Social Psychology Review* 10 (3): 252–264.

70 **"Their reportedly high":** Gerald V. O'Brien (2003). "People with cognitive disabilities: The argument from marginal cases and social work ethics." *Social Work* 48: 331–337.

70 **"a race of vermin":** Quoted in Katie Trumpener (1992). "The time of the Gypsies: A 'people without history' in the narratives of the West." *Critical Inquiry* 18 (4): 843–884.

70 **David Livingstone Smith:** David Livingstone Smith (2011). *Less Than Human: Why We Demean, Enslave, and Exterminate Others.* St. Martin's Press.

70 **"These aren't people":** See C-SPAN (@cspan) (May 16, 2018). "President Trump during California #SanctuaryCities Roundtable: 'These aren't people. These are animals.'" Tweet. https://x.com/cspan/status/996845374819192833.

70 **Experimental research:** Jacques-Philippe Leyens, Stéphanie Demoulin, Jeroen Vaes, Ruth Gaunt, and Maria Paola Paladino (2007). "Infra-humanization: The wall of group differences." *Social Issues and Policy Review* 1 (1): 139–172.

70 **One set of studies by psychologist:** Philip Atiba Goff, Jennifer L. Eberhardt, Melissa J. Williams, and Matthew Christian Jackson (2008). "Not yet human: Implicit knowledge, historical dehumanization, and contemporary consequences." *Journal of Personality and Social Psychology* 94 (2): 292–306.

71 **Decolonial theorist Aph Ko:** Animal Voices (October 18, 2019). "Aph Ko on speciesism as an extension of white supremacy." *Vancouver Co-op Radio* (podcast). https://animalvoices.org/2019/10/aph-ko-on-speciesism-as-an-extension -of-white-supremacy (around minute 43:00). See also Aph Ko (2019). *Racism as Zoological Witchcraft: A Guide to Getting Out.* Lantern.

72 **Lisa Guenther and Will Kymlicka:** Lisa Guenther (2012). "Beyond dehumanization: A post-humanist critique of solitary confinement." *Journal for Critical Animal Studies* 10 (2): 46–68; Will Kymlicka (2018). "Human rights without human supremacism." *Canadian Journal of Philosophy* 48 (6): 763–792.

72 **"we cannot fully understand":** Guenther (2012). "Beyond dehumanization" (p. 59).

72 **known as the "interspecies model of prejudice":** Kimberly Costello and Gordon Hodson (2012). "Explaining dehumanization among children: The interspecies model of prejudice." *British Journal of Social Psychology* 53 (1): 175–197.

72 **beliefs in evolution:** Stylianos Syropoulos, Uri Lifshin, Jeff Greenberg, Dylan E. Horner, and Bernhard Leidner (2022). "Bigotry and the human-animal divide: (Dis)belief in human evolution and bigoted attitudes across different cultures." *Journal of Personality and Social Psychology* 123 (6): 1264–1292.

72 **Kimberly Costello and colleagues:** Kimberly Costello and Gordon Hodson (2010). "Exploring the roots of dehumanization: The role of animal-human similarity in promoting immigrant humanization." *Group Processes and Intergroup Relations* 13 (1): 3–22; Brock Bastian, Kimberly Costello, Steve Loughnan, and Gordon Hodson (2012). "When closing the human-animal divide expands moral concern: The importance of framing." *Social Psychological and Personality Science* 3 (4): 421–429.

73 **replicated even among children:** Costello and Hodson (2012). "Explaining dehumanization among children."

CHAPTER FOUR: PRIDE THAT BLINDS

77 **Classic experiments on prosocial behavior:** For example, Joan B. Silk et al. (2005). "Chimpanzees are indifferent to the welfare of unrelated group members." *Nature* 437 (7063): 1357–1359; Claudio Tennie, Keith Jensen, and Josep Call (2016). "The nature of prosociality in chimpanzees." *Nature Communications* 7 (13915); Michael Tomasello (2014). "The ultra-social animal." *European Journal of Social Psychology* 44: 187–194.

78 **vary markedly across human cultures:** Joseph Henrich, Steven J. Heine, and Ara Norenzayan (2010). "The WEIRDest people in the world?" *Behavioral and Brain Sciences* 33: 61–83.

78 **sharing implies a contractual obligation:** See Christophe Boesch (2010). "Away from ethnocentrism and anthropocentrism: Towards a scientific understanding of 'what makes us human.'" *Behavioral and Brain Sciences* 33 (2–3): 86–87.

80 **Marine foraging behavior:** Matthew Lewis (2015). "Behavioural and isotope ecology of marine-foraging chacma baboons (*Papio ursinus*) on the Cape Peninsula, South Africa." PhD thesis, University of Cape Town. https://open.uct .ac.za/bitstream/handle/11427/15610/thesis_sci_2015_lewis_matthew _charles.pdf?sequence=1.

80 **Studies on prison inmates:** Rebecca Umbach, Adrian Raine, and Noelle R. Leonard (2018). "Cognitive decline as a result of incarceration and the effects of a CBT/MT intervention: A cluster-randomized controlled trial." *Criminal Justice and Behavior* 45 (1): 31–55.

81 **One 2018 study:** Martin Glabischnig (2018). "The value of lab and field studies for learning about the evolution of human cognition." MSc thesis, University of Amsterdam. https://doi.org/10.13140/RG.2.2.17120.74248.

81 **In another analysis:** Christophe Boesch (2007). "What makes us human (*Homo sapiens*)? The challenge of cognitive cross-species comparison." *Journal of Comparative Psychology* 121 (3): 227–240.

82 **chimps participate in highly coordinated:** Liran Samuni, Anna Preis, Tobias Deschner, Catherine Crockford, and Roman M. Wittig (2018). "Reward of la-

bor coordination and hunting success in wild chimpanzees." *Communications Biology* 1 (138): 1–9.

82 **wild chimps share not only meat:** Liran Samuni et al. (2018). "Social bonds facilitate cooperative resource sharing in wild chimpanzees." *Proceedings of the Royal Society B: Biological Sciences* 285: 20181643.

82 **majority of mutual aid:** Kevin E. Langergraber, John C. Mitani, and Linda Vigilant (2007). "The limited impact of kinship on cooperation in wild chimpanzees." *Proceedings of the National Academy of Sciences of the United States of America* 104 (19): 7786–7790.

82 **some chimps are consistently more empathic:** Christine E. Webb, Teresa Romero, Becca Franks, and Frans B. M. de Waal (2017). "Long-term consistency in chimpanzee consolation behaviour reflects empathetic personalities." *Nature Communications* 8 (292).

82 **One chimpanzee community:** Nobuyuki Kutsukake and Duncan L. Castles (2004). "Reconciliation and post-conflict third-party affiliation among wild chimpanzees in the Mahale Mountains, Tanzania." *Primates* 45 (3): 157–165.

83 **another in the Budongo Forest:** Kate Arnold and Andrew Whiten (2001). "Post-conflict behaviour of wild chimpanzees (*Pan troglodytes schweinfurthii*) in the Budongo Forest, Uganda." *Behaviour* 138 (5): 649–690.

83 **animals console bereaved group members:** Zoë Goldsborough, Edwin J. C. van Leeuwen, Kayla W. T. Kolff, Frans B. M. de Waal, and Christine E. Webb (2020). "Do chimpanzees (*Pan troglodytes*) console a bereaved mother?" *Primates* 61 (1): 93–102.

83 **coalition support to allies:** Reviewed in Christophe Boesch et al. (2019). *The Chimpanzees of the Taï Forest: 40 Years of Research.* Cambridge University Press.

84 **adopting unrelated orphans:** Boesch et al. (2019). *The Chimpanzees of the Taï Forest* (pp. 141–158).

84 **"All direct ape-human comparisons":** David A. Leavens, Kim A. Bard, and William D. Hopkins (2014). "The mismeasure of ape social cognition" (p. 491). *Animal Cognition* 22 (4): 487–504.

84 **video on a set of studies:** "Experiments with altruism in children and chimps." https://www.youtube.com/watch?v=Z-eU5xZW7cU. These videos are also available as separate downloads in the supplemental materials of the original *Science* article: Felix Warneken and Michael Tomasello (2006). "Altruistic helping in human infants and young chimpanzees." *Science* 311 (5765): 1301–1303.

85 **subtle forms of encouragement:** Note in particular how the father in the blue shirt nudges his child in the third video at https://www.youtube.com/watch?v=Z-eU5xZW7cU.

86 **"public enemy number one":** Richard Nixon Presidential Library and Museum. "President Nixon declares drug abuse 'public enemy number one.'" Accessed September 14, 2024. https://www.youtube.com/watch?v=y8TGLLQlD9M.

86 **Rat Park residents consumed:** See, for example, Bruce K. Alexander, Barry L. Beyerstein, Patricia F. Hadaway, and Robert B. Coambs (1981). "Effect of early and later colony housing on oral ingestion of morphine in rats." *Pharmacology, Biochemistry and Behavior* 15 (4): 571–576.

87 **"In both cases, the colonizers":** Bruce K. Alexander (2010). "Addiction: The view from Rat Park." Accessed September 14, 2024. https://www.brucekalex ander.com/articles-speeches/rat-park/148-addiction-the-view-from-rat-park.

87 **Stephen Jay Gould exposed:** Stephen Jay Gould (1981). *The Mismeasure of Man.* Norton.

87 **A parallel systemic blindness:** See Leavens et al. (2014). "The mismeasure of ape social cognition."

88 **psychologist Donald Hebb:** Donald O. Hebb (1947). "The effects of early experience on problem-solving at maturity." *American Psychologist* 2: 737–745.

88 **many studies have demonstrated:** Reviewed in Becca Franks (2018). "Cognition as a cause, consequence, and component of welfare." In Joy A. Mench (ed.), *Advances in Agricultural Animal Welfare* (pp. 3–24). Woodhead.

88 **wild-born chimps:** Richard K. Davenport, Charles M. Rogers, and Duane M. Rumbaugh (1973). "Long-term cognitive deficits in chimpanzees associated with early impoverished rearing." *Developmental Psychology* 9 (3): 343–347.

89 **compared with isolated calves:** Joao H. C. Costa, Marina A. G. von Keyserlingk, and Daniel M. Weary (2016). "Effects of group housing of dairy calves on behavior, cognition, performance, and health." *Journal of Dairy Science* 99 (4): 2453–2467.

89 **hens have higher well-being:** Fernanda M. Tahamtani, Janicke Nordgreen, Rebecca E. Nordquist, and Andrew M. Janczak (2015). "Early life in a barren environment adversely affects spatial cognition in laying hens (*Gallus gallus domesticus*)." *Frontiers in Veterinary Science* 2: 1–12.

89 **when cod in standard hatcheries:** Victoria A. Braithwaite and Anne G. V. Salvanes (2005). "Environmental variability in the early rearing environment generates behaviourally flexible cod: Implications for rehabilitating wild populations." *Proceedings of the Royal Society B: Biological Sciences* 272 (1568): 1107–1113.

90 **meme that went viral:** We Don't Deserve This Planet (May 17, 2020). Facebook. https://m.facebook.com/WeDontDeserveEarth/posts/716797869067104.

90 **acute challenges are good:** Michael Mendl (1999). "Performing under pressure: Stress and cognitive function." *Applied Animal Behaviour Science* 65 (3): 221–244.

91 **one 2017 study:** Stephen Ferrigno, Julian Jara-Ettinger, Steven T. Piantadosi, and Jessica F. Cantlon (2017). "Universal and uniquely human factors in spontaneous number perception." *Nature Communications* 8 (13968).

91 **physically damages the brain:** See, for instance, Bob Jacobs et al. (2022). "Putative neural consequences of captivity for elephants and cetaceans." *Reviews in*

the Neurosciences 33 (4): 439–465; Lori Marino et al. (2020). "The harmful effects of captivity and chronic stress on the well-being of orcas (*Orcinus orca*)." *Journal of Veterinary Behavior* 35: 69–82.

92 **"When Animals Lose Their Minds":** Laurel Braitman (June 9, 2014). "When animals lose their minds." *Wall Street Journal*. https://www.wsj.com/articles /when-animals-lose-their-minds-1402084124.

92 **more than 115 million animals:** Humane Society International. "About animal testing." Accessed November 17, 2022. https://www.hsi.org/news-resources /about/#.

92 **Success rates in human clinical trials:** Derek Lowe (May 9, 2019). "The latest on drug failure and approval rates." *In the Pipeline* (blog), *Science*. Accessed December 2, 2022. https://www.science.org/content/blog-post/latest-drug -failure-and-approval-rates.

92 **"every drug that fails":** Joseph P. Garner et al. (2017). "Introducing therioepistemology: The study of how knowledge is gained from animal research" (p. 103). *Lab Animal* 46 (4): 103–113.

92 **infinitesimal proportion of their average:** Garet Lahvis (2017). "Make animal models more meaningful." *Nature* 543: 623.

92 **Research on laboratory rodent health:** Jessica Cait et al. (2022). "Conventional laboratory housing increases morbidity and mortality in research rodents: Results of a meta-analysis." *BMC Biology* 20: 1–22.

93 **known as the "standardization fallacy":** Bernhard Voelkl, Hanno Würbel, Martin Krzywinski, and Naomi Altman (2021). "The standardization fallacy." *Nature Methods* 18: 3–7.

94 **a heated debate:** For details, see Lars Chittka (2022). *The Mind of a Bee* (pp. 18–20). Princeton University Press. .

94 **"By the scent":** Karl von Frisch (1971). *Bees: Their Vision, Chemical Senses, and Language* (p. 5). Cornell University Press.

95 **"all cats are gray at night":** I borrow this suggestion from Chittka (2022). *The Mind of a Bee* (p. 20).

95 **in other animals—including BIZARRE:** David A. Leavens, Kim A. Bard, and William D. Hopkins (2010). "BIZARRE chimpanzees do not represent the chimpanzee." *Behavioral and Brain Sciences* 33 (2–3): 100–101.

95 **STRANGE ("S" for "social"):** Michael M. Webster and Christian Rutz (2020). "How STRANGE are your study animals?" *Nature* 582 (7812): 337–340.

96 **"Humans are born":** Christophe Boesch (2007). "What makes us human (*Homo sapiens*)? The challenge of cognitive cross-species comparison" (p. 228). *Journal of Comparative Psychology* 121 (3): 227–240.

97 **"to predict and control":** B. F. Skinner (1953). *Science and Human Behavior* (p. 35). Macmillan.

97 **"Ornaments of all kinds":** Charles Darwin (1871/1981). *The Descent of Man, Vol. 2* (p. 86). Princeton University Press.

98 **"When the same behavior":** Eileen Crist (1998). "The ethological constitution of animals as natural objects: The technical writings of Konrad Lorenz and Nikolaas Tinbergen" (p. 84). *Biology and Philosophy* 13 (1): 61–102.

100 **point to Morgan's Canon:** Conwy Lloyd Morgan (1894). *An Introduction to Comparative Psychology.* Walter Scott. The canon states: "In no case may we interpret an action as the outcomes of the exercise of a higher psychical faculty, if it can be interpreted as the outcome of the exercise of one which stands lower in the psychological scale" (p. 53).

The lesser-known addendum to Morgan's Canon reads: "To this, however, it should be added, lest the range of the principle be misunderstood, that the canon by no means excludes the interpretation of a particular activity in terms of the higher processes if we already have independent evidence of the occurrences of these higher processes in the animal under observation" (p. 59).

In other words, if we already have evidence of mental processes in other species, then those processes should be considered valid explanations for behavior.

100 **evolutionary parsimony:** For a rich discussion of cognitive versus evolutionary parsimony, see Frans B. M. de Waal (1999). "Anthropomorphism and anthropodenial: Consistency in our thinking about humans and other animals." *Philosophical Topics* 27 (1): 255–280.

100 **"a blindness to":** de Waal (1999). "Anthropomorphism and anthropodenial" (p. 258).

101 **"to draw such a conclusion":** Brian L. Keeley (2004). "Anthropomorphism, primatomorphism, mammalomorphism: Understanding cross-species comparisons" (p. 523). *Biology and Philosophy* 19 (4): 521–540.

101 **philosophical ethologist Dominique Lestel:** See Matthew Chrulew (2014). "The philosophical ethology of Dominique Lestel." *Angelaki* 19 (3): 17–44.

102 **Food and Agriculture Organization of the United Nations:** Food and Agriculture Organization (2020). *The State of the World's Forests.* Accessed July 30, 2021. https://www.fao.org/state-of-forests.

102 **A 2019 *Science* study:** Hjalmar S. Kühl et al. (2019). "Human impact erodes chimpanzee behavioral diversity." *Science* 363 (6434): 1453–1455.

102 **appreciate that glyphosate:** María Sol Balbuena et al. (2015). "Effects of sublethal doses of glyphosate on honeybee navigation." *Journal of Experimental Biology* 218 (17): 2799–2805.

102 **disrupting the shoaling:** Ashley J. W. Ward, Alison J. Duff, Jennifer S. Horsfall, and Suzanne Currie (2008). "Scents and scents-ability: Pollution disrupts chemical social recognition and shoaling in fish." *Proceedings of the Royal Society B: Biological Sciences* 275 (1630): 101–105.

102 **That anthropogenic noise:** Graeme Shannon et al. (2016). "A synthesis of two decades of research documenting the effects of noise on wildlife." *Biological Reviews* 91 (4): 982–1005.

103 **One particularly telling study:** Alison Osbrink et al. (2021). "Traffic noise inhibits cognitive performance in a songbird." *Proceedings of the Royal Society B: Biological Sciences* 288 (1944): 20202851.

103 **Fossil-fuel combustion:** Kristopher B. Karnauskas, Shelly L. Miller, and Anna C. Schapiro (2020). "Fossil fuel combustion is driving indoor CO_2 toward levels harmful to human cognition." *GeoHealth* 4: e2019GH000237.

103 **Lead levels have:** Bruce P. Lanphear, Kim Dietrich, Peggy Auinger, and Christopher Cox (2000). "Cognitive deficits associated with blood lead concentrations <10 µg/dL in US children and adolescents." *Public Health Reports* 115 (6): 521–529.

103 **display symptoms of PTSD:** Gay A. Bradshaw, Allan N. Schore, Janine L. Brown, Joyce H. Poole, and Cynthia J. Moss (2005). "Elephant breakdown." *Nature* 433 (7028): 807.

104 **ethical indifference to their study:** See also John P. Gluck (2016). *Voracious Science and Vulnerable Animals: A Primate Scientist's Ethical Journey.* University of Chicago Press.

CHAPTER FIVE: THE MISMEASURE OF ALL THINGS

107 *This American Life*: Ira Glass (October 17, 1997). "Running after antelope." *This American Life,* episode 80. https://www.thisamericanlife.org/80/transcript.

107 **concept of the "umwelt":** Jakob von Uexküll (1992/1934). "A stroll through the worlds of animals and men: A picture book of invisible worlds." *Semiotica* 89 (4): 319–391.

108 **A 2020 study:** Mary Caswell Stoddard et al. (2020). "Wild hummingbirds discriminate nonspectral colors." *Proceedings of the National Academy of Sciences of the United States of America* 117 (26): 15112–15122.

108 **circularly polarized light:** Hanne H. Thoen, Martin J. How, Tsyr Huei Chiou, and Justin Marshall (2014). "A different form of color vision in mantis shrimp." *Science* 343 (6169): 411–413.

109 **bioinspired optical devices:** Benjamin A. Palmer et al. (2017). "The image-forming mirror in the eye of the scallop." *Science* 358 (6367): 1172–1175.

109 **five times more:** Erika Engelhaupt (April 23, 2017). "Inside the bizarre life of the star-nosed mole, world's fastest eater." *National Geographic.* https://www.nationalgeographic.com/animals/article/star-nosed-mole-touch-pain-senses.

110 **smell into account:** Sarah L. Jacobson and Joshua M. Plotnik (2020). "The importance of sensory perception in an elephant's cognitive world." *Comparative Cognition and Behavior Reviews* 15: 1–18.

110 **they can reportedly hear:** Zeena Lemon (August 22, 2006). "The barn owl can hear a mouse's heartbeat at 25 feet." *Worcester News.* https://www.worcesternews.co.uk/news/6543082.this-barn-owl-can-hear-a-mouses-heartbeat-at-25ft/.

110 **discriminates between the fishes:** Yossi Yovel and Whitlow W. L. Au (2010). "How can dolphins recognize fish according to their echoes? A statistical analysis of fish echoes." *PLOS ONE* 5 (11): 1–10.

111 **a kind of "electrotouch":** Ed Yong (2022). *An Immense World: How Animal Senses Reveal the Hidden Realms Around Us* (pp. 323–324). Random House.

111 **"the length of a moment":** Von Uexküll (1992/1934). "A stroll through the worlds of animals and men" (p. 326).

111 **waving a finger:** I borrow this suggestion from Charles Foster (2016). *Being a Beast* (p. 195). Picador.

111 **"Sloths have no right":** Quoted in Craig Holdrege (2019). "What does it mean to be a sloth?" (p. 7). Nature Institute. http://natureinstitute.org/nature/sloth .pdf.

111 **animals with fast metabolic rates:** Kevin Healy, Luke McNally, Graeme D. Ruxton, Natalie Cooper, and Andrew L. Jackson (2013). "Metabolic rate and body size are linked with perception of temporal information." *Animal Behaviour* 86 (4): 685–696.

112 **Some birds can hear:** I borrow this suggestion from Foster (2016). *Being a Beast* (p. 196).

112 **mirror self-recognition task:** Gordon G. Gallup Jr. (1970). "Chimpanzees: Self-recognition." *Science* 167: 86–87.

113 **have used a novel design:** Alexandra Horowitz (2017). "Smelling themselves: Dogs investigate their own odours longer when modified in an 'olfactory mirror' test." *Behavioural Processes* 143: 17–24; Roberto Cazzolla Gatti (2016). "Self-consciousness: Beyond the looking-glass and what dogs found there." *Ethology, Ecology and Evolution* 28 (2): 232–240.

113 **olfaction is typically neglected:** Alexandra Horowitz and Becca Franks (2020). "What smells? Gauging attention to olfaction in canine cognition research." *Animal Cognition* 23 (1): 11–18.

113 **a recent review:** Miles K. Bensky, Samuel D. Gosling, and David L. Sinn (2013). "The world from a dog's point of view: A review and synthesis of dog cognition research." *Advances in the Study of Behavior* 45: 209–406.

113 **"A dog can never":** Mary Oliver (2013). "Her grave." In *Dog Songs* (p. 23). Penguin Books.

114 **researchers in human child development:** Celia A. Brownell, Stephanie Zerwas, and Geetha B. Ramani (2007). "'So big': The development of body self-awareness in toddlers." *Child Development* 78 (5): 1426–1440.

115 **in one recent study:** Sridhar Ravi et al. (2020). "Bumblebees perceive the spatial layout of their environment in relation to their body size and form to minimize inflight collisions." *Proceedings of the National Academy of Sciences of the United States of America* 117 (49): 31494–31499.

115 **distinguish the echoes:** Eran Amichai, Gaddi Blumrosen, and Yossi Yovel (2015). "Calling louder and longer: How bats use biosonar under severe acous-

tic interference from other bats." *Proceedings of the Royal Society B: Biological Sciences* 282: 20152064.

115   **Don't dolphins have "signature whistles":** Heidi E. Harley (2008). "Whistle discrimination and categorization by the Atlantic bottlenose dolphin (*Tursiops truncatus*): A review of the signature whistle framework and a perceptual test." *Behavioural Processes* 77: 243–268.

115   **aptly titled book:** Frans de Waal (2017). *Are We Smart Enough to Know How Smart Animals Are?* Norton.

116   **"Quite simply, I was in love":** Joan Didion (1967). "Goodbye to all that." In *Slouching Towards Bethlehem* (p. 228). Farrar, Straus and Giroux.

117   **Cognitive ethology—a field:** See Donald R. Griffin (1998). "From cognition to consciousness." *Animal Cognition* 1 (1): 3–16.

117   **"Man is the measure":** Explained in Joshua J. Mark. "Protagoras of Abdera: Of all things man is the measure." *World History Encyclopedia*. Last modified January 18, 2012. https://www.worldhistory.org/article/61/protagoras-of-abdera -of-all-things-man-is-the-meas/. Protagoras's statement has traditionally been interpreted as skepticism about the possibility of attaining objective knowledge. But in the comparative context emphasized here, it implies that humans are the ultimate yardstick against which other species' abilities can be evaluated.

118   **chimpanzees could beg:** D. J. Povinelli and T. J. Eddy (1996). "Factors influencing young chimpanzees' recognition of 'attention.'" *Journal of Comparative Psychology* 110: 336–345.

118   **Brian Hare and colleagues:** Brian Hare, Josep Call, Bryan Agnetta, and Michael Tomasello (2000). "Chimpanzees know what conspecifics do and do not see." *Animal Behaviour* 59 (4): 771–785.

118   **understood what others had seen:** Brian Hare, Josep Call, and Michael Tomasello (2001). "Do chimpanzees know what conspecifics know?" *Animal Behaviour* 61 (1): 139–151.

118   **"Is Geometry a Language":** Siobhan Roberts (March 22, 2022). "Is geometry a language that only humans know?" *New York Times*. https://www.nytimes .com/2022/03/22/science/geometry-math-brain-primates.html.

119   **"intuitions of geometry":** Mathias Sablé-Meyer et al. (2021). "Sensitivity to geometric shape regularity in humans and baboons: A putative signature of human singularity." *Proceedings of the National Academy of Sciences of the United States of America* 118 (16): e2023123118.

119   **Desert ants navigate:** Matthias Wittlinger, Rudiger Wehner, and Harald Wolf (2006). "The ant odometer: Stepping on stilts and stumps." *Science* 312: 1965–1967.

119   **African lions and spotted hyenas:** Sarah Benson-Amram, Geoff Gilfillan, and Karen McComb (2017). "Numerical assessment in the wild: Insights from social carnivores." *Philosophical Transactions of the Royal Society B: Biological Sciences* 373: 20160508.

119 **Clark's nutcracker retrieves food:** Kathryn Schulz (March 29, 2021). "Why animals don't get lost." *New Yorker.* https://www.newyorker.com/magazine /2021/04/05/why-animals-dont-get-lost.

120 **homing pigeons, using precise:** Schulz (March 29, 2021). "Why animals don't get lost."

120 **"Life-forms don't line up":** Martha C. Nussbaum (2018). "Working with and for animals: Getting the theoretical framework right" (p. 5). *Journal of Human Development and Capabilities* 19 (1): 2–18.

120 **favorite cartoons, seven animals:** *Marquette Educator* (blog). Accessed January 20, 2021. https://marquetteeducator.wordpress.com/wp-content/uploads /2012/07/5232012052424iwsmt.jpeg.

121 **"Smarter Than You Think":** Jonathan Leake and Georgia Warren (January 17, 2010). "Smarter than you think." *The Times.* https://www.thetimes.co.uk /article/smarter-than-you-think-hpzhntm5hr2.

121 **"Human intelligence is":** "Of bairns and brains." (May 28, 2016). *The Economist.* https://www.economist.com/science-and-technology/2016/05/28 /of-bairns-and-brains.

121 **"Humans May Be":** Nick Longrich (October 22, 2019). "Humans may be the only intelligent life in the universe, if evolution has anything to say." *Live Science.* https://www.livescience.com/evolution-says-humans-only-intelligent -life.html.

122 **The Wikipedia entry:** "Intelligence." Wikipedia. Accessed February 10, 2021. https://en.wikipedia.org/wiki/Intelligence.

122 **A 2007 study:** Shane Legg and Marcus Hutter (2007). "A collection of definitions of intelligence." *Frontiers in Artificial Intelligence and Applications* 57: 17–24.

122 *one* **kind of intelligence:** Juliane Bräuer and colleagues labeled this restrictive perspective the "one cognition" approach, highlighting the tendency of comparative researchers to overrate human cognitive skills and to assume that certain skills cluster together in other animals as they do in humans. Juliane Bräuer, Daniel Hanus, Simone Pika, Russell Gray, and Natalie Uomini (2020). "Old and new approaches to animal cognition: There is not 'one cognition.'" *Journal of Intelligence* 8 (3): 1–25.

122 **"To a lover of music":** Nussbaum (2018). "Working with and for animals" (p. 5).

122 **describes as a "double-edged sword":** Justin Gregg (2022). *If Nietzsche Were a Narwhal.* Little, Brown.

124 **human brand of these qualities:** For instance, the zoologist Yosef Prat argues that comparative language studies often measure acoustic features that are conspicuous to humans, ignoring the unique communicative umwelt of the animal under consideration. Yosef Prat (2019). "Animals have no language, and humans are animals too." *Perspectives on Psychological Science* 14 (5): 885–893.

124 **"It is of interest":** Quoted in Jonathan Balcombe (2010). *Second Nature: The Inner Lives of Animals* (p. 83). Palgrave Macmillan.

125 **observed that forager bees:** Karl von Frisch (1993). *The Dance Language and Orientation of Bees.* Harvard University Press.

125 **one second of dancing:** BuzzAboutBees. "The honey bee dance." Accessed September 14, 2024. https://www.buzzaboutbees.net/Honey-Bee-Dance.html. Interestingly, however, different populations have different functions relating flight distance to dance duration. Fred C. Dyer (2002). "The biology of the dance language." *Annual Review of Entomology* 47: 917–949.

125 **6,864 possible combinations:** Con Slobodchikoff (2012). *Chasing Doctor Dolittle: Learning the Language of Animals* (p. 148). St. Martin's Press.

125 **skin patterns of Caribbean reef squid:** Martin Moynihan (1991). *Communication and Noncommunication by Cephalopods.* Indiana University Press; Chun Yen Lin, Yueh Chun Tsai, and Chuan Chin Chiao (2017). "Quantitative analysis of dynamic body patterning reveals the grammar of visual signals during the reproductive behavior of the oval squid *Sepioteuthis lessoniana.*" *Frontiers in Ecology and Evolution* 5 (30): 1–16.

126 **"Human language is special":** Eva Meijer (2016). "Speaking with animals: Philosophical interspecies investigations." In Morten Tønnessen, Kristin Armstrong Oma, and Silver Rattasepp (eds.), *Thinking About Animals in the Age of the Anthropocene* (p. 87). Lexington Books.

127 **"is largely determined":** Ellen Dissanayake (1988). *What Is Art For?* (p. 119). University of Washington Press.

127 **"Music is characterized":** William Forde Thompson (2009). *Music, Thought, and Feeling: Understanding the Psychology of Music* (p. 38). Oxford University Press.

127 **zoomusicologist Hollis Taylor:** Hollis Taylor (2013). "Connecting interdisciplinary dots: Songbirds, 'white rats' and human exceptionalism." *Social Science Information* 52 (2): 287–306.

127 **"everything about human music":** Iain McGilchrist (2009). *The Master and His Emissary* (p. 123). Yale University Press.

127 **Songbirds regularly sing:** Evangeline M. Rose, Nora H. Prior, and Gregory F. Ball (2022). "The singing question: Re-conceptualizing birdsong." *Biological Reviews* 97 (1): 326–342.

128 **"One could propose":** Taylor (2013). "Connecting interdisciplinary dots" (pp. 294–295).

128 **Taylor also notes:** Taylor (2013). "Connecting interdisciplinary dots" (p. 295).

128 *The Superior Human?:* *The Superior Human?* (2012). DocumentaryTube .com. https://www.documentarytube.com/videos/the-superior-human.

128 **Some have suggested:** Ted Chu (2014). *Human Purpose and Transhuman Potential.* Origin Press; Jonathan Marks (2015). *Tales of the Ex-Apes.* University of California Press.

128 **world's largest beaver dam:** Jean Thie. "The longest beaver dam in the world." EcoInformatics International. Accessed September 14, 2024. https://www.geo strategis.com/p_beavers-longestdam.htm.

129 **But termites' mounds are:** Hunter King, Samuel Ocko, and L. Mahadevan (2015). "Termite mounds harness diurnal temperature oscillations for ventilation." *Proceedings of the National Academy of Sciences of the United States of America* 112 (37): 11589–11593.

129 **termites had it figured out:** Ulrich G. Mueller, Nicole M. Gerardo, Duur K. Aanen, Diana L. Six, and Ted R. Schultz (2005). "The evolution of agriculture in insects." *Annual Review of Ecology, Evolution, and Systematics* 36: 563–595.

129 **Eastgate Centre in Zimbabwe:** Never Enough Architecture. "The Eastgate Centre." Accessed February 15, 2021. https://neverenougharchitecture.com /project/the-eastgate-centre.

130 **a feature scientists have co-opted:** Frank E. Fish, Paul W. Weber, Mark M. Murray, and Laurens E. Howle (2011). "The tubercles on humpback whales' flippers: Application of bio-inspired technology." *Integrative and Comparative Biology* 51 (1): 203–213.

130 **Shinkansen in Japan:** Jennifer Green and Anna Doble (producers) and Jules Bartl (animator). "How a kingfisher helped reshape Japan's bullet train" (video). *Thirty Animals That Made Us Smarter.* BBC. Accessed February 15, 2021. https://www.bbc.com/news/av/science-environment-47673287.

130 **Velcro was invented:** Jake Swearingen (2016). "An idea that stuck: How George de Mestral invented the Velcro brand fastener." *New York.* https://ny mag.com/vindicated/2016/11/an-idea-that-stuck-how-george-de-mestral -invented-velcro.html.

130 **"Sponges knew how":** Corydon Ireland (December 4, 2008). "Scientists explore nature's designs." *Harvard Gazette.* https://news.harvard.edu/gazette /story/2008/12/scientists-explore-natures-designs.

130 **"Backed by 300 million years":** See https://www.spintex.co.uk.

130 **Namib desert beetles:** *Wired* (September 14, 2015). "Can Namib desert beetles help us solve our drought problems?" (video). https://www.youtube.com /watch?v=TmyfqjXOf7M.

130 **Miles Davis apparently honed:** *Nova Music Blog* (October 2020). "Miles Davis." Accessed March 1, 2021. https://novamusic.blog/miles-davis.

131 **The composer Olivier Messiaen:** Rob Hudson. "Olivier Messiaen, bird song, and Carnegie Hall." Carnegie Hall. Accessed December 8, 2024. https://www .carnegiehall.org/Explore/Articles/2021/04/14/Olivier-Messiaen-Bird-Song -and-Carnegie-Hall.

131 **"is full of genius":** Henry David Thoreau (January 5, 1856, journal entry). Walden Woods Project. https://www.walden.org/wp-content/uploads/2016 /02/Journal-8-Chapter-3.pdf.

131 **the installation *MycoTunnel*:** See https://maelokko.com/Exhibitions.

132   **"They'd stand on":** David McCullough (2016). *The Wright Brothers* (p. 52). Simon and Schuster.

132   **"They would watch":** McCullough (2016). *The Wright Brothers* (p. 52).

132   **"Learning the secret of flight":** McCullough (2016). *The Wright Brothers* (p. 52).

133   **skills of non-primates "primatocentrism":** Marc Bekoff (1997). "Deep ethology, animal rights, and the Great Ape/Animal Project: Resisting speciesism and expanding the community of equals." *Journal of Agricultural and Environmental Ethics* 10 (3): 269–296.

133   **dubbed it "chimpocentrism":** Benjamin B. Beck (1982). "Chimpocentrism: Bias in cognitive ethology." *Journal of Human Evolution* 11 (1): 3–17.

134   **"common animal traits":** Thomas Suddendorf (November 21, 2013). "Are we really different from animals?" CNN. https://www.cnn.com/2013/11/21/health /animals-humans-gap.

134   **This includes popular claims:** See, for instance, American Psychological Association (August 10, 2009). "Dogs' intelligence on par with two-year-old human, canine researcher says." ScienceDaily. https://www.sciencedaily.com /releases/2009/08/090810025241.htm; Richard Gray (August 9, 2009). "Dogs as intelligent as two-year-old-children." *The Telegraph.* https://www.telegraph .co.uk/news/science/science-news/5994583/Dogs-as-intelligent-as-two-year -old-children.html.

135   **"All living beings are":** Colin Allen (2001). "Cognitive relatives and moral relations." In Benjamin B. Beck et al. (eds.), *Great Apes and Humans at an Ethical Frontier* (p. 3). Smithsonian Institution Press.

135   **Conservation research is often biased:** J. Alan Clark and Robert M. May (2002). "Taxonomic bias in conservation research." *Science* 297 (5579): 191–192.

135   **termed the "so like us" approach:** Nussbaum (2018). "Working with and for animals" (p. 3).

136   **"In short, if we line up":** Nussbaum (2018). "Working with and for animals" (p. 5).

137   **My doctoral research:** Christine E. Webb (2015). "Moving past conflict: How locomotion facilitates reconciliation in humans and chimpanzees (*Pan troglodytes*)." PhD thesis, Columbia University. https://academiccommons.colum bia.edu/doi/10.7916/D8PV6JG1.

137   **a theory paper:** Christine E. Webb, Maya Rossignac-Milon, and E. Tory Higgins (2017). "Stepping forward together: Could walking facilitate interpersonal conflict resolution?" *American Psychologist* 72 (4): 374–385.

137   **Research has since:** See, for example, Gray Atherton and Liam Cross (2020). "Walking in my shoes: Imagined synchrony improves attitudes towards outgroups." *Psychological Studies* 65 (4): 351–359.

137   **"The challenge, then, becomes":** Philip Ball (June 11, 2022). "Animal magic: Why intelligence isn't just for humans." *The Guardian.* https://www.theguard

ian.com/books/2022/jun/11/animal-magic-why-intelligence-isnt-just-for
-humans.

138 **"space of possible minds":** Aaron Sloman (1984). "The structure of the space
of possible minds." In Stephen B. Torrance and Ellis Horwood (eds.), *The Mind
and the Machine: Philosophical Aspects of Artificial Intelligence* (pp. 35–42).
Halsted Press.

138 **also known as Erwin's Law:** Rob Dunn (2021). *A Natural History of the Future.* Basic Books.

138 **with a single study in a Panamanian rainforest:** Terry L. Erwin (1982).
"Tropical forests: Their richness in Coleoptera and other arthropod species."
Coleopterists Bulletin 36 (1): 74–75. Erwin's methods were brutal; he sprayed a
fog of pesticide into the tree canopy, then collected and identified all the beetles
who fell to the forest floor.

139 **"plant blindness" that makes us:** James H. Wandersee and Elisabeth E.
Schussler (1999). "Preventing plant blindness." *American Biology Teacher* 61
(2): 82–86.

139 **"because biology is taught":** Wandersee and Schussler (1999). "Preventing
plant blindness" (p. 82).

139 **more than sixteen thousand known species of moss:** Leath Tonino (April
2016). "Two ways of knowing." *The Sun.* https://www.thesunmagazine.org
/articles/22248-two-ways-of-knowing.

139 **with every inhale:** Janine Fröhlich-Nowoisky, Daniel A. Pickersgill, Viviane
R. Després, and Ulrich Pöschl (2009). "High diversity of fungi in air particulate
matter." *Proceedings of the National Academy of Sciences of the United States of
America* 106 (31): 12814–12819.

139 **Dunn and his colleagues found:** Albert Barberán et al. (2015). "The ecology
of microscopic life in household dust." *Proceedings of the Royal Society B: Biological Sciences* 282: 20151139.

139 **Some estimate that:** Kenneth J. Locey and Jay T. Lennon (2016). "Scaling laws
predict global microbial diversity." *Proceedings of the National Academy of Sciences of the United States of America* 113 (21): 5970–5975.

139 **"Erwin's estimate had led scientists":** Dunn (2021). *A Natural History of the
Future* (p. 28).

139 **"Our perception of the world":** Dunn (2021). *A Natural History of the Future*
(p. 28).

CHAPTER SIX: THINKING OTHERWISE

141 **classic 1974 essay:** Thomas Nagel (1974). "What is it like to be a bat?" *Philosophical Review* 83 (4): 435–450.

142 **"There is nothing that we know":** David J. Chalmers (1995). "Facing up to the
problem of consciousness." *Journal of Consciousness Studies* 2 (3): 200–219.

143   *something it is like*: Though most people trace this definition of consciousness back to Nagel (1974), it appears somewhat earlier in Timothy L. S. Sprigge and Alan Montefiore (1971). "Final causes." *Aristotelian Society Supplementary* 45 (1): 149–192.

144   **led by Robert Feldman:** Robert S. Feldman, James A. Forrest, and Benjamin R. Happ (2002). "Self-presentation and verbal deception: Do self-presenters lie more?" *Basic and Applied Social Psychology* 24 (2): 163–170.

144   **"The conventional view":** Robert L. Trivers (2006). Foreword to Richard Dawkins's *The Selfish Gene* (p. 20). Oxford University Press.

144   **we lie to ourselves:** Robert Trivers (2011). *The Folly of Fools: The Logic of Deceit and Self-Deception in Human Life.* Basic Books.

145   *On a scale of 1 to 7*: The items here were adapted from the Prosocialness Scale for Adults. Gian Vittorio Caprara, Patrizia Steca, Arnaldo Zelli, and Cristina Capanna (2005). "A new scale for measuring adults' prosocialness." *European Journal of Psychological Assessment* 21 (2): 77–89.

145   **So-called subliminal stimuli:** See John A. Bargh and Tanya L. Chartrand (2014). "The mind in the middle: A practical guide to priming and automaticity research." In Harry T. Reis and Charles M. Judd (eds.), *Handbook of Research Methods in Social and Personality Psychology* (pp. 311–344). Cambridge University Press.

145   **human behavior expresses:** Sara J. Shettleworth (2010). "Clever animals and killjoy explanations in comparative psychology." *Trends in Cognitive Science* 14: 477–481.

145   **third "narcissistic wound":** Sigmund Freud (1917/1955). "A difficulty in the path of psycho-analysis." In James Strachey (ed.), *The Complete Psychological Works of Sigmund Freud, Vol. 17* (pp. 137–144). Hogarth.

146   **much deeper role:** George Lakoff and Mark Johnson (1999). *Philosophy in the Flesh.* Basic Books.

146   **human memory is less reliable:** Armin Schnider (2008). *The Confabulating Mind: How the Brain Creates Reality.* Oxford University Press.

146   **anthropofabulation:** Cameron Buckner (2013). "Morgan's canon, meet Hume's dictum: Avoiding anthropofabulation in cross-species comparisons." *Biology and Philosophy* 28 (5): 853–871.

147   **nonverbal cues are:** Judith A. Hall, Terrence G. Horgan, and Nora A. Murphy (2019). "Nonverbal communication." *Annual Review of Psychology* 70: 271–294.

147   **"Jij en ik . . .":** Quoted in Harry Wels (2013). "Whispering empathy: Transdisciplinary reflections on research methodology." In Bert Musschenga and Anton van Harskamp (eds.), *What Makes Us Moral?* (p. 158). Springer Science.

148   **"The tendency to demand":** Donald R. Griffin (1998). "From cognition to consciousness" (p. 13). *Animal Cognition* 1 (1): 3–16.

149   **"'Can they suffer?'":** Jeremy Bentham (1789/1948). *An Introduction to the Principles of Morals and Legislation.* Hafner Press.

149 **Cambridge Declaration on Consciousness:** Philip Low et al. (2012). "The Cambridge Declaration on Consciousness." Francis Crick Memorial Conference on Consciousness in Human and Non-Human Animals. https://fcmcon ference.org/img/CambridgeDeclarationOnConsciousness.pdf.

151 **"a relic of attempts":** Donald M. Broom (2001). "The evolution of pain." *Flemish Veterinary Journal* 70: 17–21.

151 **injected trouts' lips:** Lynne U. Sneddon, Victoria A. Braithwaite, and Michael J. Gentle (2003). "Do fishes have nociceptors? Evidence for the evolution of a vertebrate sensory system." *Proceedings of the Royal Society B: Biological Sciences* 270 (1520): 1115–1121.

151 **responses are ameliorated:** For a review, see Katherine A. Sloman, Ian A. Bouyoucos, Edward J. Brooks, and Lynne U. Sneddon (2019). "Ethical considerations in fish research." *Journal of Fish Biology* 94 (4): 556–577.

151 **invertebrates like crustaceans:** For a review, see Robert W. Elwood (2019). "Discrimination between nociceptive reflexes and more complex responses consistent with pain in crustaceans." *Philosophical Transactions of the Royal Society B: Biological Sciences* 374: 20190368.

152 **crustaceans don't have this ability:** Animals Australia. "6 incredible facts that will change the way you think about lobsters." Accessed May 2, 2021. https://animalsaustralia.org/latest-news/6-incredible-lobster-facts.

152 **across a large part:** Robyn J. Crook, Roger T. Hanlon, and Edgar T. Walters (2013). "Squid have nociceptors that display widespread long-term sensitization and spontaneous activity after bodily injury." *Journal of Neuroscience* 33 (24): 10021–10026.

153 **growing evidence to suggest they will be:** Matilda Gibbons, Andrew Crump, Meghan Barrett, Sajedeh Sarlak, Jonathan Birch, and Lars Chittka (2022). "Can insects feel pain? A review of the neural and behavioural evidence." *Advances in Insect Physiology* 63: 155–229.

153 **"There may be extraordinary":** Charles Darwin (1871/1981). *The Descent of Man, Vol. 1* (p. 145). Princeton University Press.

153 **are exquisitely complex:** Lars Chittka and Jeremy Niven (2009). "Are bigger brains better?" *Current Biology* 19 (21): R995–R1008.

153 **philosopher Jeff Sebo points out:** Jeff Sebo (July 27, 2021). "Don't farm bugs." *Aeon.* https://aeon.co/essays/on-the-torment-of-insect-minds-and-our-moral-duty -not-to-farm-them.

154 **Nearly every gene:** Kay Prüfer et al. (2012). "The bonobo genome compared with the chimpanzee and human genomes." *Nature* 486 (7404): 527–531.

154 **Our kind has fewer genes:** Alexander Werth (2012). "Avoiding the pitfall of progress and associated perils of evolutionary education." *Evolution: Education and Outreach* 5 (2): 249–265.

155 **"the most complex object":** Quoted in *Science Friday* (June 14, 2013). "Decoding 'the most complex object in the known universe.'" NPR. https://www

.npr.org/2013/06/14/191614360/decoding-the-most-complex-object-in-the-universe.

155 **"the most complicated organization":** Isaac Asimov (1986). Foreword (p. xv) in Judith Hooper and Dick Teresi. *The Three-Pound Universe*. Macmillan.

155 **brains of parrots and songbirds:** Seweryn Olkowicz et al. (2016). "Birds have primate-like numbers of neurons in the forebrain." *Proceedings of the National Academy of Sciences of the United States of America* 113 (26): 7255–7260.

156 **linearly scaled-up primate brain:** Suzana Herculano-Houzel (2012). "The remarkable, yet not extraordinary, human brain as a scaled-up primate brain and its associated cost." *Proceedings of the National Academy of Sciences of the United States of America* 109 (1): 10661–10668; Suzana Herculano-Houzel (2009). "The human brain in numbers: A linearly scaled-up primate brain." *Frontiers in Human Neuroscience* 3 (31): 1–11.

156 **are less remarkable:** See, for example, Robert A. Barton and Chris Venditti (2013). "Human frontal lobes are not relatively large." *Proceedings of the National Academy of Sciences of the United States of America* 110 (22): 9001–9006; Ralph L. Holloway (2002). "How much larger is the relative volume of area 10 of the prefrontal cortex in humans?" *American Journal of Physical Anthropology* 118 (4): 399–401.

At best, we stand out in the relative size of the cerebral cortex as a percentage of brain mass, but not by much. The human cerebral cortex is the largest among mammals in its relative size, at 75.5 percent, 75.7 percent, or even 84 percent (depending on the study) of the entire brain mass of volume. But other animals are not far off: the cerebral cortex represents 73 percent of the entire brain mass of chimpanzees, 73.4 percent in the short-finned whale, and 74.5 percent in the horse. See Herculano-Houzel (2012). "The remarkable, yet not extraordinary, human brain" (p. 10661).

156 **"The human brain is so unique":** Brian Resnick (May 23, 2018). "Why do humans have such huge brains? Scientists have a few hypotheses." *Vox.* https://www.vox.com/science-and-health/2018/5/23/17377200/human-brain-size-evolution-nature.

156 **roughly $12 billion:** Matej Mikulic (May 16, 2024). "Total neuroscience funding by the NIH from FY 2013 to FY 2025." Statista. https://www.statista.com/statistics/712866/total-neuroscience-funding-by-the-national-institutes-for-health/.

156 **"If it seems":** Gary Marcus (April 22, 2013). "The mystery of human uniqueness." *Nautilus.* https://nautil.us/the-mystery-of-human-uniqueness-234309.

157 **avian pallium is nucleated:** Erich D. Jarvis et al. (2005). "Avian brain and vertebrate brain evolution." *Nature Reviews Neuroscience* 6 (2): 151–159.

158 **"everything we don't really understand":** Quoted in Ferris Jabr (November 7, 2012). "How brainless slime molds redefine intelligence." *Scientific American.* https://www.scientificamerican.com/article/brainless-slime-molds.

158   **pattern of the cities:** Atsushi Tero et al. (2010). "Rules for biologically inspired adaptive network design." *Science* 327 (22): 439–442.

159   **took only twice as long:** Liping Zhu, Song Ju Kim, Masahiko Hara, and Masashi Aono (2018). "Remarkable problem-solving ability of unicellular amoeboid organism and its mechanism." *Royal Society Open Science* 5: 180396.

159   **Evidence is mounting:** For a helpful primer to this area of research, see Michael Pollan (December 15, 2013). "The intelligent plant." *New Yorker.* https://www.newyorker.com/magazine/2013/12/23/the-intelligent-plant.

160   **"Trees do not have":** Quoted in Richard Grant (March 2018). "Do trees talk to each other?" *Smithsonian.* https://www.smithsonianmag.com/science-nature/the-whispering-trees-180968084.

160   **an influential book:** Peter Tompkins and Christopher Bird (1973). *The Secret Life of Plants.* Harper and Row.

160   **series of studies:** Monica Gagliano, Michael Renton, Martial Depczynski, and Stefano Mancuso (2014). "Experience teaches plants to learn faster and forget slower in environments where it matters." *Oecologia* 175 (1): 63–72.

161   **they quickly fold:** NikTheCat (January 7, 2008). "Mimosa pudica—the sensitive plant." https://www.youtube.com/watch?v=BLTcVNyOhUc&t=20s.

161   **"a logical statement":** Derrick Jensen (2016). *The Myth of Human Supremacy* (p. 36). Seven Stories Press.

162   **"are thus revealed as endless":** James Bridle (2022). *Ways of Being* (p. 76). Farrar, Straus and Giroux.

163   **release oils and chemicals:** Heidi M. Appel and Rex B. Cocroft (2014). "Plants respond to leaf vibrations caused by insect herbivore chewing." *Oecologia* 175 (4): 1257–1266.

163   **lima bean plants:** Marcel Dicke et al. (1990). "Isolation and identification of volatile kairomone that affects acarine predator-prey interactions." *Journal of Chemical Ecology* 16 (2): 381–396.

163   **chemical messages that warn:** Richard Karban, Louie H. Yang, and Kyle F. Edwards (2014). "Volatile communication between plants that affects herbivory: A meta-analysis." *Ecology Letters* 17 (1): 44–52.

163   **fungal "wood wide web":** Monika A. Gorzelak, Amanda K. Asay, Brian J. Pickles, and Suzanne W. Simard (2015). "Inter-plant communication through mycorrhizal networks mediates complex adaptive behaviour in plant communities." *AoB Plants* 7: plv050.

163   **from other plant species:** See, for example, Satoru Sukegawa et al. (2018). "Pest management using mint volatiles to elicit resistance in soy: Mechanism and application potential." *Plant Journal* 96 (5): 910–920.

163   **single root apex can detect:** Amy Fleming (April 5, 2020). "The secret life of plants: How they memorise, communicate, problem solve and socialize." *The Guardian.* https://www.theguardian.com/environment/2020/apr/05/smarty-plants-are-our-vegetable-cousins-more-intelligent-than-we-realise.

163 **recognize self from nonself:** Michal Gruntman and Ariel Novoplansky (2004). "Physiologically mediated self/non-self discrimination in roots." *Proceedings of the National Academy of Sciences of the United States of America* 101 (11): 3863–3867.

164 **searocket plants:** Mudra V. Bhatt, Aditi Khandelwal, and Susan A. Dudley (2011). "Kin recognition, not competitive interactions, predicts root allocation in young *Cakile edentula* seedling pairs." *New Phytologist* 189 (4): 1135–1142.

164 **networks facilitate recovery:** Yuan Yuan Song, Suzanne W. Simard, Allan Carroll, William W. Mohn, and Ren Sen Zeng (2015). "Defoliation of interior Douglas-fir elicits carbon transfer and stress signalling to ponderosa pine neighbors through ectomycorrhizal networks." *Scientific Reports* 5: 1–9.

164 **dosed-up *Mimosa* plants:** Ken Yokawa et al. (2018). "Anaesthetics stop diverse plant organ movements, affect endocytic vesicle recycling and ROS homeostasis, and block action potentials in Venus flytraps." *Annals of Botany* 122 (5): 747–756.

164 **compounds that are anesthetic:** František Baluška and Ken Yokawa (2021). "Anaesthetics and plants: From sensory systems to cognition-based adaptive behaviour." *Protoplasma* 258: 449–454.

164 **"They're living organisms":** Quoted in JoAnna Klein (February 2, 2018). "Sedate a plant, and it seems to lose consciousness. Is it conscious?" *New York Times.* https://www.nytimes.com/2018/02/02/science/plants-consciousness-anesthe sia.html.

164 **"If we would embrace":** Quoted in Leath Tonino (April 2016). "Two ways of knowing." *The Sun.* https://www.thesunmagazine.org/issues/484/two-ways-of -knowing.

165 *The Power of Movement in Plants:* Charles Darwin and Francis Darwin (1880). *The Power of Movement in Plants.* John Murray.

165 **root-brain hypothesis:** František Baluška, Stefano Mancuso, Dieter Volkmann, and Peter Barlow (2009). "The 'root-brain' hypothesis of Charles and Francis Darwin." *Plant Signaling and Behavior* 4 (12): 1121–1127.

166 **the first to admit:** Paco Calvo, Monica Gagliano, Gustavo M. Souza, and Anthony Trewavas (2020). "Plants are intelligent, here's how." *Annals of Botany* 125 (1): 11–28.

166 **"The answer, unreservedly, is 'no'":** Devang Mehta (February 14, 2018). "Plants are not conscious, whether you can 'sedate' them or not." Massive Science. https:// massivesci.com/articles/plants-conscious-intelligence-movement-sedate.

166 **"No brain, no pain":** Quoted in Pollan (2013). "The intelligent plant." For a sense of the opposition, see also Lincoln Taiz et al. (2019). "Plants neither possess nor require consciousness." *Trends in Plant Science* 24 (8): 677–687.

167 **"Perception, memory, valence":** Pamela Lyon (October 21, 2012). "On the origin of minds." *Aeon.* https://aeon.co/essays/the-study-of-the-mind-needs-a -copernican-shift-in-perspective.

168   **known as panpsychism:** See Philip Goff (2019). *Galileo's Error: Foundations for a New Science of Consciousness.* Pantheon Books.

168   **the sentience of the more-than-human world:** This phrase was coined by David Abram (1996). *The Spell of the Sensuous.* Vintage Books.

CHAPTER SEVEN: RELATIONSHIP MATTERS

173   **In a landmark paper:** Barbara Smuts (2001). "Encounters with animal minds." *Journal of Consciousness Studies* 8 (5–7): 293–309.

175   **Japanese word *kyokan*:** Masao Kawai (1969). *Nihonzaru no Seitai (Ecology of Japanese Monkeys).* Kawade Shobo Shinsha. See also Pamela J. Asquith (1996). "Japanese science and Western hegemonies: Primatology and the limits set to questions." In Laura Nader (ed.), *Naked Science: Anthropological Inquiry into Boundaries, Power, and Knowledge* (pp. 239–256). Routledge.

175   **"You can't study men":** C. S. Lewis (1946). *That Hideous Strength* (pp. 70–71). Macmillan.

177   **In one longitudinal study:** Christine E. Webb, Becca Franks, Teresa Romero, E. Tory Higgins, and Frans B. M. de Waal (2014). "Individual differences in chimpanzee reconciliation relate to social switching behaviour." *Animal Behaviour* 90: 57–63.

177   **striking personality differences:** Christine E. Webb, Teresa Romero, Becca Franks, and Frans B. M. de Waal (2017). "Long-term consistency in chimpanzee consolation behaviour reflects empathetic personalities." *Nature Communications* 8 (292).

177   **chimpanzees even express stress:** Zoë Goldsborough, Elisabeth H. M. Sterck, Frans B. M. de Waal, and Christine E. Webb (2022). "Individual variation in chimpanzee (*Pan troglodytes*) repertoires of abnormal behaviour." *Animal Welfare* 31 (1): 125–135.

177   **"My awareness of":** Smuts (2001). "Encounters with animal minds" (p. 301).

178   **"I have no doubt":** Len Howard (1952). *Birds as Individuals* (p. 18). Collins.

178   **"Humans who are over-conceited":** Len Howard (1956). *Living with Birds* (p. 127). Collins.

179   **"There was no trouble":** Howard (1952). *Birds as Individuals* (p. 149).

180   **"It is a significant fact":** Charles Darwin (1871/1981). *The Descent of Man, Vol. 1* (p. 46). Princeton University Press.

180   **philosopher Mary Midgley:** Mary Midgley (2001). "Being objective." *Nature* 410 (6830): 753.

181   **it confounds "objectivity":** I borrow this suggestion from Becca Franks. See https://www.watr-lab.org/values.

181   **colleagues and I call "the empathy taboo":** Christine Webb, Becca Franks, Monica Gagliano, and Barbara Smuts (2023). "Un-tabooing empathy: The benefits of empathic science with nonhuman research participants." In Francesca

Mezzenzana and Daniela Peluso (eds.), *Conversations on Empathy: Interdisciplinary Perspectives on Imagination and Radical Othering* (pp. 216–234). Routledge.

181 **Cartesian and behaviorist ideologies:** The assumption that some purely objective standpoint is even possible is traceable to Descartes's well-known separation of the immaterial human mind (or subject) from the material, mechanical world of nature, including the body (or objects). Later, the school of behaviorism would also emphasize detachment from the entities and events being observed, lest engaging alter animal behavior or prompt one to make unscientific interpretations of animals' lives.

181 **"Here, what we do not know":** Stanley Cavell (1976). *Must We Mean What We Say? A Book of Essays* (p. 69). Cambridge University Press.

182 **rats and mice behaved differently:** Robert E. Sorge et al. (2014). "Olfactory exposure to males, including men, causes stress and related analgesia in rodents." *Nature Methods* 11: 629–632.

182 **Mogil disclosed in an interview:** Quoted in McGill Newsroom (April 28, 2014). "The scent of a man." https://www.mcgill.ca/newsroom/channels/news/scent-man-235492.

182 **one 2022 study found:** Polymnia Georgiou et al. (2022). "Experimenters' sex modulates mouse behaviors and neural responses to ketamine via corticotropin releasing factor." *Nature Neuroscience* 25 (9): 1191–1200.

182 **"Beings do not pre-exist":** Donna Haraway (2003). *The Companion Species Manifesto* (p. 6). Prickly Paradigm Press.

182 **relationship between scientist and subject:** Studies on overimitation—the tendency to copy a demonstrator's actions that are not necessary to achieve a goal—also reveal why human-animal relationships should be taken more seriously. It has been claimed that overimitation is a uniquely human capacity, thought to play a key role in explaining why human culture can accumulate over time. According to the philosopher Kristen Andrews, however, claims that other species don't engage in overimitation are based on studies that don't take into account the relationship between observer and subject. While chimpanzees don't overimitate in experiments, there is evidence that they do overimitate in-group members (Tetsuro Matsuzawa, who has a lifetime research relationship with the chimpanzee Ai, found that she overimitated his irrelevant tool use). Supporting this relational interpretation, recent studies find that dogs overimitate their caregivers but not unknown researchers. See Kristin Andrews (2020). *How to Study Animal Minds*. Cambridge University Press; Ludwig Huber, Kaja Salobir, Roger Mundry, and Giulia Cimarelli (2020). "Selective overimitation in dogs." *Learning and Behavior* 48 (1): 113–123.

183 **"If the eighth day":** Oskar Pfungst (1911). *Clever Hans* (p. 22). Henry Holt.

184 **"merely a motor reaction":** Pfungst (1911). *Clever Hans* (p. 221).

184 **"*no manner of intelligence*":** Pfungst (1911). *Clever Hans* (p. 79).

184 **How ironic that one of:** Two scholars encourage us to rethink the legacy of this notorious scandal: Eileen Crist (1997). "From questions to stimuli, from answers to reactions: The case of Clever Hans." *Semiotica* 113 (1–2): 1–42; Vinciane Despret (2015). "Who made Clever Hans stupid?" (trans. Matthew Chrulew). *Angelaki* 20 (2): 77–85.

185 **"The understandings we derive":** Arnold Arluke and Clinton R. Sanders (1996). *Regarding Animals* (p. 78). Temple University Press.

185 **"A face, a signature":** Maurice Merleau-Ponty (1945/2002). *Phenomenology of Perception* (p. 67). Routledge.

186 **the writer Melanie Challenger argues:** Melanie Challenger (2021). *How to Be Animal*. Penguin.

187 **counterpart in modern physics:** For example, Carlo Rovelli (1996). "Relational quantum mechanics." *International Journal of Theoretical Physics* 35 (8): 1637–1678.

187 **a new *shared* language:** This notion is explored further in Stuart G. Shanker and Barbara J. King (2002). "The emergence of a new paradigm in ape language research." *Behavioral and Brain Sciences* 25 (5): 646–656.

188 **with other species as "strange kinship":** Maurice Merleau-Ponty (2003). *Nature* (p. 214). Northwestern University Press.

188 **"shared embodiment in a shared world":** Kelly Oliver (2007). "Stopping the anthropological machine: Agamben with Heidegger and Merleau-Ponty" (p. 18). *PhaenEx* 2 (2): 1–23.

189 **"What Does a Parrot Know":** Charles Siebert (January 28, 2016). "What does a parrot know about PTSD?" *New York Times*. https://www.nytimes.com/2016/01/31/magazine/what-does-a-parrot-know-about-ptsd.html.

189 **Clinical psychologist Olga Solomon:** Olga Solomon (2015). "'But—he'll fall!': Children with autism, interspecies intersubjectivity, and the problem of 'being social.'" *Culture, Medicine, and Psychiatry* 39: 323–344.

189 **aquariums had improved outcomes:** Nancy E. Edwards and Alan M. Beck (2002). "Animal-assisted therapy and nutrition in Alzheimer's disease." *Western Journal of Nursing Research* 24 (6): 697–712.

190 **"It is difficult to escape":** James A. Serpell (2010). "Animal-assisted interventions in historical perspective." In Aubrey H. Fine (ed.), *Handbook on Animal-Assisted Therapy* (3rd ed., p. 29). Elsevier.

190 **shown promise in treating:** For reviews, see Fabrizio Bert et al. (2016). "Animal assisted intervention: A systematic review of benefits and risks." *European Journal of Integrative Medicine* 8: 695–706; Janelle Nimer and Brad Lundahl (2007). "Animal-assisted therapy: A meta-analysis." *Anthrozoös* 20 (3): 225–238.

190 **four hundred million companion animals:** The Animal Health Institute Primer (2022). "The economic and social contributions of the animal health industry" (p. 2). https://ahi.org/wp-content/uploads/AHI-Primer-December-2022-Final-w-Infographic.pdf.

190 **more than two-thirds of all households:** The Animal Health Institute Primer (2022). "The economic and social contributions of the animal health industry" (p. 3).

190 **more than 90 percent of those:** The Animal Health Institute Primer (2022). "The economic and social contributions of the animal health industry" (p. 3).

190 **Studies consistently show:** The Animal Health Institute Primer (2022). "The economic and social contributions of the animal health industry" (p. 17).

191 **accelerates the healing process:** See Kaitlyn Gillis and Birgitta Gatersleben (2015). "A review of psychological literature on the health and wellbeing benefits of biophilic design." *Buildings* 5 (3): 948–963; Roger S. Ulrich (1993). "Biophilia, biophobia, and natural landscapes." In Stephen R. Kellert and Edward O. Wilson (eds.), *The Biophilia Hypothesis* (pp. 73–137). Island Press.

191 **formerly incarcerated individuals:** Megan Holmes and Tina M. Waliczek (2019). "The effect of horticultural community service programs on recidivism." *HortTechnology* 29 (4): 490–495.

191 **humans have a "need to belong":** Kelly Ann Allen, DeLeon L. Gray, Roy F. Baumeister, and Mark R. Leary (2022). "The need to belong: A deep dive into the origins, implications, and future of a foundational construct." *Educational Psychology Review* 34 (2): 1133–1156.

191 **reveal a "ripple effect":** Lisa J. Wood, Billie Giles-Corti, Max K. Bulsara, and Darcy A. Bosch (2007). "More than a furry companion: The ripple effect of companion animals on neighborhood interactions and sense of community." *Society and Animals* 15: 43–56.

191 **sense of neighborhood stability:** Art McCabe (2014). "Community gardens to fight urban youth crime and stabilize neighborhoods." *International Journal of Child Health and Human Development* 7 (3): 1–14.

192 **"slow looking" exercise:** The concept of "slow looking" is explored in depth in Shari Tishman (2017). *Slow Looking: The Art and Practice of Learning Through Observation.* Routledge.

193 **In his 1923 book:** Martin Buber (1923/1970). *I and Thou* (trans. Walter Kaufmann). Scribner.

194 **"All real living is meeting":** Buber (1923/1970). *I and Thou* (p. 11).

CHAPTER EIGHT: THE ENTANGLED BANK

196 **"when we try to pick out":** John Muir (1988/1911). *My First Summer in the Sierra* (p. 110). Sierra Club Books.

197 **increase in sociosexual:** Jake S. Brooker, Christine E. Webb, Edwin J. C. van Leeuwen, Stephanie Kordon, Frans B. M. de Waal, and Zanna Clay (2025). "Bonobos and chimpanzees overlap in sexual behaviour patterns during social tension." *Royal Society Open Science* 12 (3): 242031.

197 **sometimes called "nature, red in tooth and claw":** This expression is taken from Alfred Lord Tennyson's 1950 poem "In Memoriam A. H. H."

198 **selfish genes don't necessarily make:** Quoted in *Faith and Reason*. "Richard Dawkins interview transcript." PBS. Accessed September 14, 2024. https://www.pbs.org/faithandreason/transcript/dawk-frame.html.

199 **"I use the term":** Charles Darwin (1859/1902). *On the Origin of Species* (p. 101). American Home Library.

199 **an entangled bank:** Darwin (1859/1902). *On the Origin of Species* (p. 315).

199 **"It will have been":** Charles Darwin (1871/1981). *The Descent of Man, Vol. 1* (p. 82). Princeton University Press.

199 **"Don't compete!":** Peter Kropotkin (1902). *Mutual Aid: A Factor of Evolution* (p. 62). Knopf.

200 **variation *within* the two species:** Jake S. Brooker, Christine E. Webb, Frans B. M. de Waal, and Zanna Clay (2024). "The expression of empathy in human's closest living relatives, bonobos and chimpanzees: Current and future directions." *Biological Reviews* 99 (4): 1556–1575.

A similar picture is emerging in the wild. In a recent study comparing rates of male aggression in three wild bonobo communities at the Kokolopori Bonobo Reserve, Democratic Republic of Congo, and two chimpanzee communities at Gombe National Park, Tanzania, researchers found that Kokolopori bonobos showed higher overall rates of male-male aggression than Gombe chimpanzees. See Maud Mouginot, Michael L. Wilson, Nisarg Desai, and Martin Surbeck (2024). "Differences in expression of male aggression between wild bonobos and chimpanzees." *Current Biology* 34: 1780–1785.

201 **Suzanne Simard, a professor:** See her moving memoir: Suzanne Simard (2021). *Finding the Mother Tree.* Knopf.

202 **cover of *Nature*:** Suzanne W. Simard, David A. Perry, Melanie D. Jones, David D. Myrold, Daniel M. Durrall, and Randy Molina (1997). "Net transfer of carbon between ectomycorrhizal tree species in the field." *Nature* 388: 579–582.

203 **protecting the plants against pests:** See, for example, Yuan Yuan Song, Suzanne W. Simard, Allan Carroll, William W. Mohn, and Ren Sen Zeng (2015). "Defoliation of interior Douglas-fir elicits carbon transfer and stress signalling to ponderosa pine neighbors through ectomycorrhizal networks." *Scientific Reports* 5: 1–9.

203 **"The old foresters":** Quoted in Ferris Jabr (December 2, 2020). "The social life of forests." *New York Times.* https://www.nytimes.com/interactive/2020/12/02/magazine/tree-communication-mycorrhiza.html.

203 **Skepticism about Simard's research:** See, for example, Justine Karst, Melanie D. Jones, and Jason D. Hoeksema (2023). "Positive citation bias and overinterpreted results lead to misinformation on common mycorrhizal networks in forests." *Nature Ecology and Evolution* 7 (4): 501–511.

For an article summarizing the key arguments, see Gabriel Popkin (November 7, 2022). "Are trees talking underground? For scientists, it's in dispute." *New York Times.* https://www.nytimes.com/2022/11/07/science/trees-fungi-talking.html.

203 **"We don't ask":** Quoted in Richard Grant (March 2018). "Do trees talk to each other?" *Smithsonian.* https://www.smithsonianmag.com/science-nature/the-whispering-trees-180968084.

204 **"in nature we never":** Quoted in James Wood (1893). *Dictionary of Quotations from Ancient and Modern, English and Foreign Sources* (p. 188). Warne.

204 **"is interaction and reciprocal":** Quoted in Andrea Wulf (2015). *The Invention of Nature* (p. 59). John Murray.

204 **"the subversive science":** Paul Shepard and Daniel McKinley (eds.) (1969). *The Subversive Science: Essays Toward an Ecology of Man.* Houghton Mifflin. See also Paul Sears (1964). "Ecology: A subversive subject." *BioScience* 14: 11–13.

204 **notable offshoot is deep ecology:** Arne Naess (1973). "The shallow and the deep, long-range ecology movement." *Inquiry* 16 (1–4): 95–100.

204 **Influential thinkers have cautioned:** See, for example, Timothy Morton (2007). *Ecology Without Nature.* Harvard University Press; Bruno Latour (1993). *We Have Never Been Modern.* Harvard University Press.

205 **Strathern's studies of the Hagen:** Marilyn Strathern (1980). "No nature, no culture: The Hagen case." In Carol MacCormack and Marilyn Strathern (eds.), *Nature, Culture and Gender* (pp. 174–222). Cambridge University Press.

205 **the word "dividual":** Marilyn Strathern (1988). *The Gender of the Gift.* University of California Press.

205 **a fungus and an alga:** Newer research suggests a third partner (a type of yeast) may be involved in constituting lichen. Toby Spribille et al. (2016). "Basidiomycete yeasts in the cortex of ascomycete macrolichens." *Science* 353 (6298): 488–492.

206 **the term "symbiosis":** Jan Sapp (1994). *Evolution by Association: A History of Symbiosis* (p. 6). Oxford University Press.

206 **known as endosymbiosis:** Lynn Margulis (1981). *Symbiosis in Cell Evolution.* W. H. Freeman.

207 **"The view of evolution":** Lynn Margulis (1986). *Microcosmos: Four Billion Years of Evolution from Our Microbial Ancestors* (pp. 28–29). Summit Books.

207 **initially scoffed and laughed at:** For a sense of the adversity Margulis encountered, see John Feldman (director) (2017). *Symbiotic Earth: How Lynn Margulis Rocked the Boat and Started a Scientific Revolution.* Bullfrog Films; Lynn Margulis (1995). "Gaia is a tough bitch." In John Brockman (ed.), *The Third Culture: Beyond the Scientific Revolution* (pp. 129–146). Simon & Schuster.

207 **a new term: "involution":** Carla Hustak and Natasha Myers (2012). "Involutionary momentum: Affective ecologies and the sciences of plant/insect encounters." *Differences* 23 (3): 74–118.

208   **"rolling, curling, turning inwards":** Quoted in Hustak and Myers (2012). "Involutionary momentum" (p. 96).

208   **We're enmeshed in a wide:** Various scholars have thus proposed other modes of relationality that are nonhierarchical and not arborescent, such as the rhizome formulated by Gilles Deleuze and Félix Guattari. Gilles Deleuze and Félix Guattari (1987). *A Thousand Plateaus* (trans. Brian Massumi). University of Minnesota Press.

208   **"We are human only":** David Abram (1996). *The Spell of the Sensuous* (p. 22). Vintage Books.

209   **bacteria in your gut:** Ed Yong (2016). *I Contain Multitudes* (p. 8). Harper-Collins.

209   **The number in your mouth:** Lynn Margulis and Dorion Sagan (2007). *Dazzle Gradually: Reflections on the Nature of Nature* (p. 34). Chelsea Green.

209   **about three pounds:** Rebecca Jacobson (April 23, 2014). "Can we save our body's ecosystem from extinction?" PBS News. https://www.pbs.org/newshour/science/theres-extinction-happening-stomach.

209   **ability to solve complex memory:** Gabrielle L. Davidson, Amy C. Cooke, Crystal N. Johnson, and John L. Quinn (2018). "The gut microbiome as a driver of individual variation in cognition and functional behaviour." *Philosophical Transactions of the Royal Society B: Biological Sciences* 373 (1756).

209   **interventions that alter:** John F. Cryan and Timothy G. Dinan (2012). "Mind-altering microorganisms: The impact of the gut microbiota on brain and behaviour." *Nature Reviews Neuroscience* 13 (10): 701–712.

209   **cows themselves can't eat grass:** I borrow this suggestion from Merlin Sheldrake (2020). *Entangled Life* (p. 91). Penguin.

209   **90 percent of the bacterial species:** Yuichi Hongoh (2010). "Diversity and genomes of uncultured microbial symbionts in the termite gut." *Bioscience, Biotechnology and Biochemistry* 74 (6): 1145–1151.

210   **lineages also disappear:** James T. Staley (1997). "Biodiversity: Are microbial species threatened?" *Current Opinion in Biotechnology* 8 (3): 340–345.

210   **we are technological devices:** John Gray's observation was inspired by Lynn Margulis and Dorion Sagan's writing. John Gray (2002). *Straw Dogs* (p. 16). Farrar, Straus and Giroux.

210   **"Most organisms are bacteria":** Myra J. Hird (2010). "Meeting with the microcosmos" (p. 37). *Environment and Planning D: Society and Space* 28 (1): 36–39.

211   **"We are all lichens":** Scott F. Gilbert, Jan Sapp, and Alfred I. Tauber (2012). "A symbiotic view of life: We have never been individuals." *Quarterly Review of Biology* 87 (4): 325–341.

211   **titled "The Year 3011":** Dan Piraro (June 28, 2011). "The Year 3011." Bizarro Comics.com, distributed by King Features. https://funnyjunk.com/3011/sdiuLfq/.

212 **"Why are humans so successful":** Gary Stix (September 1, 2014). "What makes humans different than any other species?" *Scientific American.* https://www.scientificamerican.com/article/what-makes-humans-different-than-any-other-species.

213 **"Humans have certainly":** Luiz Villazon. "If the human race was wiped out, which species would dominate?" *BBC Science Focus.* Accessed September 14, 2024. https://www.sciencefocus.com/science/if-the-human-race-was-wiped-out-which-species-would-dominate.

213 **mosses have thrived:** Krista Tippet interview with Robin Wall Kimmerer (February 25, 2016). "The intelligence of plants." *On Being with Krista Tippett* (podcast). https://onbeing.org/programs/robin-wall-kimmerer-the-intelligence-of-plants-2022.

214 **as Quentin Atkinson:** Quentin D. Atkinson and Jennifer Jacquet (2022). "Challenging the idea that humans are not designed to solve climate change." *Perspectives on Psychological Science* 17 (3): 619–630.

215 **defy the parameters:** For a deeper discussion of this phenomenon, see Yael Wyner and Rob DeSalle (2020). "Distinguishing extinction and natural selection in the Anthropocene: Preventing the panda paradox through practical education measures." *BioEssays* 42: 1900206.

216 **journalist Elizabeth Kolbert:** Elizabeth Kolbert (2021). *Under a White Sky.* Crown.

217 **"Nature is not only":** Frank E. Egler (1970). *The Way of Science* (p. 21). Hafner.

217 **human technology frequently struggles:** This idea (along with examples) is carefully explored in Rob Dunn (2021). *A Natural History of the Future.* Basic Books.

218 **called human *exempt*ionalism:** Riley E. Dunlap and William R. Catton (1994). "Toward an ecological sociology: The development, current status, and probable future of environmental sociology." In William Vincent D'Antonio, Masamichi S. Sasaki, and Yoshio Yonebayashi (eds.), *Ecology, Society and the Quality of Social Life* (pp. 11–31). Transaction Publishers.

218 **our species' distinct ingenuity:** See, for example, John Asafu-Adjaye et al. (2015). "An ecomodernist manifesto." Accessed September 14, 2024. http://www.ecomodernism.org/manifesto-english.

219 **problems growing trees:** Anupum Pant. "The role of wind in a tree's life." *Awesci.* Accessed December 16, 2024. https://awesci.com/the-role-of-wind-in-a-trees-life/#google_vignette.

220 **"But man is a part of nature":** Quoted in Linda J. Lear (1997). *Rachel Carson: Witness for Nature* (p. 450). Henry Holt.

220 **Deborah Bird Rose:** Deborah Bird Rose (2006). "What if the angel of history were a dog?" *Cultural Studies Review* 12 (1): 67–78.

221 **"the body below the neck":** Derek Parfit (2012). "We are not human beings." *Philosophy* 87 (1): 5–28.

221   **Melanie Challenger argues that these ambitions:** Melanie Challenger (2021). *How to Be Animal*. Penguin.

222   **cultural anthropologist Ernest Becker:** Ernest Becker (1973). *The Denial of Death*. Free Press.

222   **need to proclaim that "I am not an animal":** Reviewed in Lori Marino and Michael Mountain (2015). "Denial of death and the relationship between humans and other animals." *Anthrozoös* 28 (1): 5–21.

222   **one 2001 study:** Jamie L. Goldenberg et al. (2001). "I am not an animal: Mortality salience, disgust, and the denial of human creatureliness." *Journal of Experimental Psychology* 130 (3): 427–435.

222   **consumer marketing context:** Alexander Davidson and Michel Laroche (2018). "Consumer preferences for human uniqueness in marketing communications." *Journal of Marketing Communications* 24 (5): 506–517.

223   ***Welwitschia mirabilis* plant:** Ian Sample (May 2, 2010). "The oldest living organisms: Ancient survivors with a fragile future." *The Guardian*. https://www.theguardian.com/theobserver/2010/may/02/rachel-sussman-oldest-plants.

223   **Old Tjikko, the nickname given:** Patrik Qvist (June 18, 2014). "Deep time: Finding Old Tjikko." Dark Mountain Project. https://dark-mountain.net/deep-time-finding-old-tjikko.

223   **Lichens can be:** Rachel Sargent Mirus (February 15, 2021). "Lichens: Winter survivalists." *Northern Woodlands*. https://northernwoodlands.org/outside_story/article/lichens-winter.

223   **Wood frogs can remain:** "Biological miracle." National Park Service. Accessed September 14, 2023. https://www.nps.gov/gaar/learn/nature/wood-frog-page-2.htm.

223   **And *Turritopsis dohrnii*:** American Museum of Natural History (May 4, 2015). "The immortal jellyfish." Accessed September 15, 2023. https://www.amnh.org/explore/news-blogs/on-exhibit-posts/the-immortal-jellyfish.

224   **"One cannot appropriate":** Tim Ingold (1986). *The Appropriation of Nature: Essays on Human Ecology and Social Relations* (p. 135). Manchester University Press.

224   **"because we are part":** Robert Bringhurst and Jan Zwicky (2018). *Learning to Die: Wisdom in the Age of Climate Crisis* (p. 13). University of Regina Press.

CHAPTER NINE: OUR INDIGENOUS INHERITANCE

225   **"The essential role":** Jean Clottes (2016). *What Is Paleolithic Art?* (p. 143). University of Chicago Press.

For more on the marginality of human figures in Paleolithic cave paintings, see Barbara Ehrenreich (December 12, 2019). "'Humans were not centre stage': How ancient cave art puts us in our place." *The Guardian*. https://www

.theguardian.com/artanddesign/2019/dec/12/humans-were-not-centre-stage -ancient-cave-art-painting-lascaux-chauvet-altamira.

226 **Indigenous knowledge emphasizes "mutual reciprocity":** Gregory Cajete (2000). *Native Science* (p. 79). Clear Light Publishers.

226 **"Everything is related":** Cajete (2000). *Native Science* (p. 75).

227 **"The Western world, long rooted":** George E. Tinker (2004). "The stones shall cry out: Consciousness, rocks, and Indians" (p. 106). *Wičazo Ša Review* 19 (2): 105–125.

227 **bethany ojalehto and colleagues:** bethany l. ojalehto, Douglas L. Medin, William S. Horton, Salino G. Garcia, and Estefano G. Kays (2015). "Seeing cooperation or competition: Ecological interactions in cultural perspectives." *Topics in Cognitive Science* 7 (4): 624–645.

228 **"the world is inhabited":** Eduardo Viveiros de Castro (1998). "Cosmological deixis and Amerindian perspectivism" (p. 469). *Journal of the Royal Anthropological Institute* 4 (3): 469–488.

228 **"Animals are people":** Viveiros de Castro (1998). "Cosmological deixis and Amerindian perspectivism" (p. 470).

228 **what humans see:** Viveiros de Castro (1998). "Cosmological deixis and Amerindian perspectivism" (pp. 477–478); The full quote reads: "It could only be this way, since, being people in their own sphere, non-humans see things as 'people' do. But the things that they see are different: what to us is blood, is maize beer to the jaguar; what to the souls of the dead is a rotting corpse, to us is soaking manioc; what we see as a muddy waterhole, the tapirs see as a great ceremonial house."

228 **This is not to impose a "human" quality:** To deem this anthropomorphism would be to assume the primacy of the human quality. That is, it is not "anthropomorphic" to compare human and animal in this way, because they share a common existential status, namely as living beings or persons, even if they have different natures. See Tim Ingold (2000). "Hunting and gathering as ways of perceiving the environment." In *The Perception of the Environment* (pp. 40–60). Routledge.

228 **"all beings, and not just humans":** Eduardo Kohn (2013). *How Forests Think* (p. 132). University of California Press.

229 **pejorative and racist manner:** See Edward B. Tylor (1871). *Primitive Culture*. Bretano. On p. 417, Tylor asserts, "Animism characterizes tribes very low in the scale of humanity, and thence ascends, deeply modified in its transmission, but from first to last preserving an unbroken continuity, into the midst of high modern culture."

229 **"Animists are people who":** Graham Harvey (2005). *Animism: Respecting the Living World* (p. xi). Columbia University Press.

229 **"The greatest peril":** As recounted by Aua, an Iglulik Inuit shaman, to the Danish ethnographer Knud Rasmussen, whose text is likely the source from which most popular versions of the aphorism derive. Knud Rasmussen (1979).

Intellectual Culture of the Iglulik Eskimos: Report of the Fifth Thule Expedition, Vol. 7 (p. 56). AMS Press.

229 **Colin Scott's research:** Colin Scott (1989). "Knowledge construction among the Cree hunters: Metaphors and literal understanding." *Journal de la Société des Américanistes* 75: 193–208. See Alison Leigh Lilly's blog for additional insights on Cree hunting rituals: Alison Leigh Lilly (January 30, 2014). "Anthropocentrism and animal instinct." *Holy Wild Blog.* https://alisonleighlilly.com /2014/01/30/anthropocentrism-and-animal-instinct.

229 **"somehow appropriate to a culture":** Robin Wall Kimmerer (2012). "Learning the grammar of animacy" (p. 6). *Leopold Outlook* (Winter): 1–9.

230 **roughly 30 percent:** Kimmerer (2012). "Learning the grammar of animacy" (p. 6).

230 **"Science is a language of distance":** Kimmerer (2012). "Learning the grammar of animacy" (p. 5).

231 **Indigenous African communities knew:** See NYU Primatology (@nyuprimatology) tweet (July 14, 2020). "'The Dari is almost human. This creature grabs an amount of palm nuts in its hand and with a stone of the other hand he breaks them and eats them' Duarte Pacheco Pereira (1508)." https://twitter .com/nyuprimatology/status/1283060889588031491.

231 **"there is an intricate":** Quoted in Suzanne Simard (2021). *Finding the Mother Tree* (p. 283). Knopf.

231 **Métis scholar Zoe Todd:** Zoe Todd (2016). "An indigenous feminist's take on the ontological turn: 'Ontology' is just another word for colonialism." *Journal of Historical Sociology* 29 (1): 4–22.

231 **Many posthumanist "realizations":** For other examples, see Juanita Sundberg (2014). "Decolonizing posthumanist geographies." *Cultural Geographies* 21 (1): 33–47.

232 **"The moral status of animals is a longstanding":** Christine E. Webb, Peter Woodford, and Elise Huchard (2019). "Animal ethics and behavior science: An overdue discussion?" *BioScience* 69 (10): 778–788.

232 **"Some moderns, both scientists":** Gregory Cajete (2000). *Native Science* (p. 3). Clear Light Publishers.

232 **Leroy Little Bear:** Leroy Little Bear (2000). Foreword to Cajete (2000), *Native Science.*

232 **Ojibwe scholar Megan Bang:** Megan Bang, Ananda Marin, and Douglas Medin (2018). "If Indigenous peoples stand with the sciences, will scientists stand with us?" *Daedalus* 147 (2): 148–159.

233 **fifty-four separate sections:** Bang, Marin, and Medin (2018). "If Indigenous peoples stand with the sciences" (p. 150).

233 **Two chimpanzee males:** Detailed in Jake S. Brooker, Christine E. Webb, and Zanna Clay (2020). "Fellatio among male sanctuary-living chimpanzees." *Behaviour* 158 (1): 77–87.

233 **in a larger study:** Aaron A. Sandel and Rachna B. Reddy (2021). "Sociosexual behaviour in wild chimpanzees occurs in variable contexts and is frequent between same-sex partners." *Behaviour* 158 (3–4): 249–276.

233 **more than 1,500:** José M. Gómez, Adela Gónzalez-Megías, and Miguel Verdú (2023). "The evolution of same-sex sexual behaviour in mammals." *Nature Communications* 14 (5719).

233 **alternative hypotheses for how same-sex:** Julia D. Monk, Erin Giglio, Ambika Kamath, Max R. Lambert, and Caitlin E. McDonough (2019). "An alternative hypothesis for the evolution of same-sex sexual behaviour in animals." *Nature Ecology and Evolution* 3 (12): 1622–1631.

233 **adaptive role in maintaining:** Gómez, Gónzalez-Megías, and Verdú (2023). "The evolution of same-sex sexual behaviour in mammals."

233 **A 2024 survey:** Karyn Anderson et al. (2024). "Same-sex sexual behaviour among mammals is widely observed, yet seldomly reported: Evidence from an online expert survey." *PLOS ONE* 19 (6): e0304885.

234 **prehistoric humans depicted:** Gabor Horvath, Etelka Farkas, Ildiko Boncz, Miklos Blaho, and Gyorgy Kriska (2012). "Cavemen were better at depicting quadruped walking than modern artists: Erroneous walking illustrations in the fine arts from prehistory to today." *PLOS ONE* 7 (12): e49786.

235 **"They've invented everything":** Quoted in Judith Thurman (June 16, 2008). "First impressions." *New Yorker*. https://www.newyorker.com/magazine/2008 /06/23/first-impressions.

235 **animal tracking was the original science:** Louis Liebenberg (1990). *The Art of Tracking: The Origin of Science*. David Philip Publishers.

235 **"Tracking involves intense concentration":** Louis Liebenberg (2006). "Persistence hunting by modern hunter-gatherers" (p. 1024). *Current Anthropology* 47 (6): 1017–1026.

235 **"must *become* that animal":** Liebenberg (1990). *The Art of Tracking* (p. 95).

235 **"perception gained from using":** Cajete (2000). *Native Science* (p. 2).

236 **The event, titled "Friendship":** Kincentric Leadership (2023). *Friendship with the More Than Human World*. https://www.kincentricleadership.org/event-info /friendship-with-the-more-than-human-world.

236 **"intellectual twin to science":** Vine Deloria Jr. (1995). *Red Earth, White Lies*. Harper and Row.

237 **does not mean *blending*:** See Leath Tonino (April 2016). "Two ways of knowing." *The Sun*. https://www.thesunmagazine.org/issues/484/two-ways-of -knowing.

237 **"learning to see":** Cheryl Bartlett, Murdena Marshall, and Albert Marshall (2012). "Two-Eyed Seeing and other lessons learned within a co-learning journey of bringing together Indigenous and mainstream knowledges and ways of knowing" (p. 335). *Journal of Environmental Studies and Sciences* 2 (4): 331–340.

237 **a team of Western scientists:** Aisling Rayne et al. (2020). "Centring Indigenous knowledge systems to re-imagine conservation translocations." *People and Nature* 2 (3): 512–526.

237 **Kivalliq Wildlife Board:** Crown-Indigenous Relations and Northern Affairs Canada (December 9, 2022*).* "Kivalliq Wildlife Board uses Inuit knowledge and Western science to study the impact of climate change on food security." https://www.canada.ca/en/crown-indigenous-relations-northern -affairs/news/2022/12/kivalliq-wildlife-board-uses-inuit-knowledge-and -western-science-to-study-the-impact-of-climate-change-on-food-security .html.

238 **sustainable agriculture is now aiding:** Marcus Vinícius C. Schmidt et al. (2021). "Indigenous knowledge and forest succession management in the Brazilian Amazon: Contributions to reforestation of degraded areas." *Frontiers in Forests and Global Change* 4: 605925.

238 **by ascribing agency:** Thanks to Dr. Roger Dube for this insight.

238 **significant challenges remain:** See, for instance, Jayalaxshmi Mistry and Andrea Berardi (2016). "Bridging Indigenous and scientific knowledge." *Science* 352: 26–27; Saima May Sidik et al. (2022). "Weaving Indigenous knowledge into the scientific method." *Nature* 601: 285–287.

238 **calls this concept Place-Thought:** Vanessa Watts (2013). "Indigenous place-thought and agency amongst humans and non-humans (First Woman and Sky Woman go on a European world tour!)." *Decolonization: Indigeneity, Education and Society* 2 (1): 20–34.

239 **"Place-Thought is":** Watts (2013). "Indigenous place-thought and agency amongst humans and non-humans" (p. 21).

239 **"not of making":** Tim Ingold (2000). *The Perception of the Environment* (p. 42). Routledge.

239 **attended a lecture:** Bathsheba Demuth (November 6, 2019). "Do whales judge us? Interspecies history and ethics." Harvard University, Environment Forum at the Mahindra Humanities Center. https://www.youtube.com/watch?v=yh _kJA0Naug.

240 **The Yupik refer to this latter action as *angyi*:** Bathsheba Demuth (October 29, 2019). "Turn and live with animals." *Aeon.* https://aeon.co/essays/the-act -of-giving-and-the-chance-of-life-on-a-finite-planet.

241 **"personhood" is not determined:** See Graham Harvey (2017). "If not all stones are alive . . .: Radical relationality in animism studies." *Journal for the Study of Religion, Nature and Culture* 4: 481–497.

241 **"When an Indigenous":** Watts (2013). "Indigenous place-thought and agency amongst humans and non-humans" (p. 32).

242 **"Our understandings of the world":** Watts (2013). "Indigenous place-thought and agency amongst humans and non-humans" (p. 22).

242 **Eve Tuck and K. Wayne Yang:** Eve Tuck and K. Wayne Yang (2012). "Decolonization is not a metaphor." *Decolonization: Indigeneity, Education and Society* 1 (1): 1–40.

242 **updated Ecuadorian Constitution:** Republic of Ecuador (2008). Constitution of 2008, Chapter 7, Article 71. Accessed at https://pdba.georgetown.edu/Constitutions/Ecuador/english08.html.

242 **Vilcabamba River's rights:** See Erin Daly (2012). "The Ecuadorian exemplar: The first ever vindications of constitutional rights of nature." *Review of European Community and International Environmental Law* 21: 63–66.

243 **"all the rights":** New Zealand Parliamentary Counsel Office. Te Urewera Act 2014, Section 1.3.11. https://www.legislation.govt.nz/act/public/2014/0051/latest/whole.html#DLM6183705.

243 **Courts in Bolivia, Colombia, and India:** Mari Margil (May 23, 2018). "Our laws make slaves of nature. It's not just humans who need rights." *The Guardian.* https://www.theguardian.com/commentisfree/2018/may/23/laws-slaves-nature-humans-rights-environment-amazon.

243 **law professor Christopher Stone:** Christopher D. Stone (1972). "Should trees have standing? Towards legal rights for natural objects." *Southern California Law Review* 45: 450–501. https://iseethics.wordpress.com/wp-content/uploads/2013/02/stone-christopher-d-should-trees-have-standing.pdf.

243 **"a bit unthinkable":** Stone (1972). "Should trees have standing?" (p. 453).

244 **"where 'recognition' is":** Glen S. Coulthard (2014). *Red Skin, White Masks: Rejecting the Colonial Politics of Recognition* (pp. 30–31). University of Minnesota Press.

244 **Indigenous peoples manage:** Stephen T. Garnett et al. (2018). "A spatial overview of the global importance of Indigenous lands for conservation." *Nature Sustainability* 1 (7): 369–374.

245 **Research reveals that annual:** Peter Veit and Helen Ding (October 7, 2016). "Protecting Indigenous land rights makes good economic sense." World Resources Institute. https://www.wri.org/insights/protecting-indigenous-land-rights-makes-good-economic-sense.

245 **It is estimated:** World Bank (April 6, 2023). "Indigenous peoples." https://www.worldbank.org/en/topic/indigenouspeoples.

245 **Notably, biodiversity levels:** See, for instance, Richard Schuster, Ryan R. Germain, Joseph R. Bennett, Nicholas J. Reo, and Peter Arcese (2019). "Vertebrate biodiversity on indigenous-managed lands in Australia, Brazil, and Canada equals that in protected areas." *Environmental Science and Policy* 101: 1–6.

245 **Spatial analyses indicate:** Alejandro Estrada et al. (2022). "Global importance of Indigenous Peoples, their lands, and knowledge systems for saving the world's primates from extinction." *Science Advances* 8: eabn2927.

245   **"Indigenous peoples sustain":** Steve Nitah (2021). "Indigenous peoples proven to sustain biodiversity and address climate change: Now it's time to recognize and support this leadership" (p. 907). *One Earth* 4: 907–909.

245   **The World Bank estimates:** World Bank (April 6, 2023). "Indigenous peoples." https://www.worldbank.org/en/topic/indigenouspeoples.

246   **Indigenous languages are being lost:** Daniel Nettle and Suzanne Romaine (2002). *Vanishing Voices.* Oxford University Press.

246   **Studies suggest that similar forces:** Larry J. Gorenflo, Suzanne Romaine, Russell A. Mittermeier, and Kristen Walker-Painemilla (2012). "Co-occurrence of linguistic and biological diversity in biodiversity hotspots and high biodiversity wilderness areas." *Proceedings of the National Academy of Sciences of the United States of America* 109 (21): 8032–8037.

246   **the basis of the "overkill hypothesis":** Paul S. Martin (1989). "Prehistoric overkill: A global model." In Paul S. Martin and Richard G. Klein (eds.), *Quaternary Extinctions* (pp. 354–404). University of Arizona Press.

246   **these extinctions remain hotly debated:** For a sense of the opposition, see David J. Meltzer (2015). "Pleistocene overkill and North American mammalian extinctions." *Annual Review of Anthropology* 44 (1): 33–53.

246   **article titles claim "Humans Caused":** Geomar (January 24, 2017). "Humans caused megafauna extinction in Australia." https://www.geomar.de /news/article?tx_news_pi1%5baction%5d=detail&tx_news_pi1%5bcon troller%5d=News&tx_news_pi1%5bactbackPid%5d=12123&tx_news_pi1 %5bbackPid%5d=12123&tx_news_pi1%5bnews%5d=4961.

246   **"Global Late Quaternary":** Christopher Sandom, Søren Faurby, Brody Sandel, and Jens Christian Svenning (2014). "Global late Quaternary megafauna extinctions linked to humans, not climate change." *Proceedings of the Royal Society B: Biological Sciences* 281: 20133254.

246   **"Humans Did Not Drive":** Scott Hocknull et al. (May 19, 2020). "Humans did not drive Australia's megafauna to extinction—climate change did." *The Guardian.* https://www.theguardian.com/science/2020/may/19/humans -australia-megafauna-to-extinction-climate-queensland.

246   **"Climate Change, Not Human":** Mathew Stewart, W. Christopher Carleton, and Huw S. Groucutt (2021). "Climate change, not human population growth, correlates with Late Quaternary megafauna declines in North America." *Nature Communications* 12 (965).

247   **"Anyone apprised of the palaeolithic":** Quoted in Paul Kingsnorth and George Monbiot (August 17, 2009). "Is there any point in fighting to stave off industrial apocalypse?" *The Guardian.* https://www.theguardian.com/com mentisfree/cif-green/2009/aug/17/environment-climate-change.

247   **"40,000 Years of Extinction":** Paul S. Martin (1990). "40,000 years of extinctions on the 'planet of doom.'" *Palaeogeography, Palaeoclimatology, Palaeoecology* 82: 187–201. I thank David B. Lauterwasser's blog for pointing me to these

useful examples: David B. Lauterwasser (October 20, 2023). "Pleistocene over-kill!" *An Anamist's Ramblings*. https://animistsramblings.substack.com/p/pleis tocene-overkill.

247 **"fundamentally what man is"**: Martin Heidegger (1977). "The age of the world picture" (p. 153). In William Lovitt (ed.), *The Question Concerning Technology, and Other Essays*. Harper Torchbooks.

247 **"the younger brothers"**: Robin Wall Kimmerer (2013). *Braiding Sweetgrass* (p. 9). Milkweed Editions.

248 **if we condense:** See Greenpeace (@greenpeace) Tumblr post (August 1, 2024). "The Earth is 4.6 billion years old. Scaling to 46 years, humans have been here 4 hours, the industrial revolution began 1 minute ago, and in that time we've destroyed more than half the world's forests." https://greenpeaceusa.tumblr .com/post/93508666790/the-earth-is-46-billion-years-old-scaling-to-46.

249 **"already inhabit what our ancestors"**: Kyle Powys Whyte (2017). "Our ancestors' dystopia now: Indigenous conservation and the Anthropocene" (p. 207). In Ursula Heise, Jon Christensen, and Michelle Niemann (eds.), *Routledge Companion to the Environmental Humanities*. Routledge.

249 **be found throughout Western history:** For overviews, see David Skrbina (2007). *Panpsychism in the West*. MIT Press; Mary Jane Rubenstein (2018). *Pantheologies: Gods, Worlds, Monsters*. Columbia University Press; Wouter J. Hanegraaff (2013). *Western Esotericism*. Bloomsbury Academic.

250 **"The word 'animism'":** Cajete (2000). *Native Science* (p. 27).

250 **relates, is "remember to remember":** Robin Wall Kimmerer. "Returning the gift." Grateful Living. Accessed September 14, 2024. https://grateful.org/re source/returning-the-gift.

CHAPTER TEN: COMING DOWN TO EARTH

253 **Psychologist Dacher Keltner:** See Dacher Keltner (2023). *Awe: The Transformative Power of Everyday Wonder*. Penguin Press.

253 **"is experienced as being much larger":** Dacher Keltner and Jonathan Haidt (2003). "Approaching awe, a moral, spiritual, and aesthetic emotion." *Cognition and Emotion* 17 (2): 297–314.

253 **"a chaos of delight":** Charles Darwin (1832). *Beagle Diary* (1831–1836). Darwin Online. Accessed July 10, 2024. https://darwin-online.org.uk/converted /manuscripts/Darwin_C_R_BeagleDiary_EHBeagleDiary.html.

253 **many people consider:** See, for instance, Michelle N. Shiota, Dacher Keltner, and Amanda Mossman (2007). "The nature of awe: Elicitors, appraisals, and effects on self-concept." *Cognition and Emotion* 21 (5): 944–963.

254 **include a collective mindset:** Yang Bai et al. (2017). "Awe, the diminished self, and collective engagement: Universals and cultural variations in the small self." *Journal of Personality and Social Psychology* 113 (2): 185–209.

254   **heightened prosocial behavior:** Paul K. Piff, Pia Dietze, Matthew Feinberg, Daniel M. Stancato, and Dacher Keltner (2015). "Awe, the small self, and prosocial behavior." *Journal of Personality and Social Psychology* 108 (6): 883–899.

254   **reduction in self-focus:** Shiota, Keltner, and Mossman (2007). "The nature of awe."

254   **researchers call "the small self":** Piff et al. (2015). "Awe, the small self, and prosocial behavior."

254   **awe fosters humility:** Jennifer E. Stellar et al. (2018). "Awe and humility." *Journal of Personality and Social Psychology* 114 (2): 258–269.

256   **"Unquestioned beliefs are the real":** Derrick Jensen (2016). *The Myth of Human Supremacy* (p. 16). Seven Stories Press.
        Jensen credits the idea to Robert Combs (1978). *Vision of the Voyage* (p. 2). Memphis State University Press.

258   **In "The New Story":** Thomas Berry (2003). "The new story." In Arthur Fabel and Donald St. John (eds.), *Teilhard in the 21st Century: The Emerging Spirit of Earth* (pp. 77–88). Orbis Books.

258   **Albert Einstein reputedly:** Rockefeller Foundation (October 3, 2014). "Defining the problem to find the solution." *Insights.* https://www.rockefellerfounda tion.org/insights/perspective/defining-problem-find-solution.

259   **"Even though this paradigm":** Berry (2003). "The new story" (p. 85).

259   **The Mother Tree Project:** https://mothertreeproject.org.

260   **I.N.S.E.C.T. Wall Twin:** Dan Parker et al. (2023). "I.N.S.E.C.T. Wall Twin: Designing for and with insects, fungi, and humans." *Temes de Disseny* 39: 228–247.

260   **"do-nothing farming":** See Masanobu Fukuoka (1985). *The Natural Way of Farming.* Japan Publications.

261   **"fallen in love outward":** Robinson Jeffers (1988). "The tower beyond tragedy." In Tim Hunt (ed.), *The Collected Poetry of Robinson Jeffers* (p. 178). Stanford University Press. I borrow this suggestion from David Abram (1996). *The Spell of the Sensuous* (p. 271). Vintage Books.

262   **Philosophers Valerie Tiberius:** Valerie Tiberius and John D. Walker (1998). "Arrogance." *American Philosophical Quarterly* 35 (4): 379–390.

262   **"mirror onto the soul":** Tiberius and Walker (1998). "Arrogance" (p. 387).

263   **As Abram points out:** David Abram (July 22, 2012). "On being human in a more-than-human world." Center for Humans and Nature. https://human sandnature.org/to-be-human-david-abram.

263   **Humility involves holding:** Joseph Chancellor and Sonja Lyubomirsky (2013). "Humble beginnings: Current trends, state perspectives, and hallmarks of humility." *Social and Personality Psychology Compass* 7 (11): 819–833.

263   **most understudied virtues:** June Price Tangney (2000). "Humility: Theoretical perspectives, empirical findings and directions for future research." *Journal of Social and Clinical Psychology* 19 (1): 70–82.

Index